Wireless Communications

Wireless Communications

Brett Rice

MURPHY & MOORE
www.murphy-moorepublishing.com

Published by Murphy & Moore Publishing
1 Rockefeller Plaza,
New York City, NY 10020, USA
www.murphy-moorepublishing.com

Wireless Communications
Brett Rice

International Standard Book Number: 978-1-63987-571-9 (Hardback)

Cataloging-in-Publication Data

Wireless communications / Brett Rice.
p. cm.
Includes bibliographical references and index.
ISBN 978-1-63987-571-9
1. Wireless communication systems. 2. Mobile communication systems.
3. Telecommunication systems. 4. Telecommunication. I. Rice, Brett.
TK5103.2 .W57 2022
621.384--dc23

Table of Contents

Preface

This book aims to help a broader range of students by exploring a wide variety of significant topics related to this discipline. It will help students in achieving a higher level of understanding of the subject and excel in their respective fields. This book would not have been possible without the unwavered support of my senior professors who took out the time to provide me feedback and help me with the process. I would also like to thank my family for their patience and support.

The transfer of signals between two or more points that are not connected by wires or cables is known as wireless communications. It is also known as over the air communication. It focuses on transmitting information through some form of energy. This form of communication can be achieved through radio, sonic, free space optical and electromagnetic induction. This field can be used to transfer information over to short distances using Bluetooth and even to longer distances using deep space radio communications. Some examples of equipment's that use this technology are global positioning system units, wireless headsets and headphones, cordless telephones, radio navigation, satellite navigation etc. This book elucidates the concepts and innovative models around prospective developments with respect to this field. While understanding the long-term perspectives of the topics, the book makes an effort in highlighting their impact as a modern tool for the growth of the discipline. As this field is emerging at a rapid pace, the contents of this book will help the readers understand the modern concepts and applications of wireless communication.

A brief overview of the book contents is provided below:

Chapter - Introduction

The method of transmission and communication of information from one point to another without the use of physical mediums like wires and cables is called wireless communication. Satellite communication, near-field communication, mobile telecommunications, indoor and satellite navigation systems, etc. fall under its domain. This is an introductory chapter which will briefly introduce wireless communications.

Chapter - Wireless Network and its Types

Wireless network uses radio waves to establish communication between computers and other network devices. A few of its types include wireless PAN, wireless ad hoc network, wireless LAN, wireless WAN, and global area network. This chapter sheds light on wireless network and its types for a thorough understanding of the subject.

Chapter – Wireless Transmission and Technologies

Wireless transmission converts the digital data into wireless signals and spreads within its frequency range through the use of antennas. Radio transmission, infrared transmission, Bluetooth, Wi-Fi and microwave transmission are a few modes of transmission. This chapter has been carefully written to provide an easy understanding of wireless transmission and technologies.

Chapter – Analog Communication and Modulation

Analog communication uses continuous signals and modulation for transmission of data such as voice, image, video, etc. Some of its concepts are amplitude modulation, single-sideband modulation, space modulation, angle modulation, etc. All these concepts related to analog communication have been carefully analyzed in this chapter.

Chapter – Digital Communication and Modulation

The mode of communication in which the information is encoded in a digital format and transferred electronically is known as digital communication. Amplitude shift keying, frequency shift keying, minimum shift keying, phase shift keying, etc. are some of the techniques used in digital transmission and modulation. This chapter discusses the related aspects of digital communication in detail.

Brett Rice

1

Introduction

The method of transmission and communication of information from one point to another without the use of physical mediums like wires and cables is called wireless communication. Satellite communication, near-field communication, mobile telecommunications, indoor and satellite navigation systems, etc. fall under its domain. This is an introductory chapter which will briefly introduce wireless communications.

Wireless Communications

Wireless communication involves the transmission of information over a distance without the help of wires, cables or any other forms of electrical conductors.

Wireless communication is a broad term that incorporates all procedures and forms of connecting and communicating between two or more devices using a wireless signal through wireless communication technologies and devices.

Features of Wireless Communication

The evolution of wireless technology has brought much advancement with its effective features.

- The transmitted distance can be anywhere between a few meters (for example, a television's remote control) and thousands of kilometers (for example, radio communication).
- Wireless communication can be used for cellular telephony, wireless access to the internet, wireless home networking, and so on.
- Other examples of applications of radio wireless technology include GPS units, garage door openers, wireless computer mice, keyboards and headsets, headphones, radio receivers, satellite television, broadcast television and cordless telephones.

Advantages to Wireless Communication

Wireless communication involves transfer of information without any physical connection between two or more points. Because of this absence of any 'physical infrastructure', wireless communication has certain advantages. This would often include collapsing distance or space. Wireless communication has several advantages; the most important ones are discussed below:

Cost Effectiveness

Wired communication entails the use of connection wires. In wireless networks, communication does not require elaborate physical infrastructure or maintenance practices. Hence the cost is reduced.

Example – Any company providing wireless communication services does not incur a lot of costs, and as a result, it is able to charge cheaply with regard to its customer fees.

Flexibility

Wireless communication enables people to communicate regardless of their location. It is not necessary to be in an office or some telephone booth in order to pass and receive messages.

Miners in the outback can rely on satellite phones to call their loved ones, and thus, help improve their general welfare by keeping them in touch with the people who mean the most to them.

Convenience

Wireless communication devices like mobile phones are quite simple and therefore allow anyone to use them, wherever they may be. There is no need to physically connect anything in order to receive or pass messages.

Example – Wireless communications services can also be seen in Internet technologies such as Wi-Fi. With no network cables hampering movement, we can now connect with almost anyone, anywhere, anytime.

Speed

Improvements can also be seen in speed. The network connectivity or the accessibility were much improved in accuracy and speed.

Example – A wireless remote can operate a system faster than a wired one. The wireless control of a machine can easily stop its working if something goes wrong, whereas direct operation can't act so fast.

Accessibility

The wireless technology helps easy accessibility as the remote areas where ground lines can't be properly laid, are being easily connected to the network.

Example – In rural regions, online education is now possible. Educators no longer need to travel to far-flung areas to teach their lessons.

Constant Connectivity

Constant connectivity also ensures that people can respond to emergencies relatively quickly.

Example – A wireless mobile can ensure you a constant connectivity though you move from place to place or while you travel, whereas a wired land line can't.

The different types of wireless communication mainly include, IR wireless communication, satellite communication, broadcast radio, Microwave radio, Bluetooth, Zigbee etc.

Satellite Communication

Satellite communication is one type of self contained wireless communication technology, it is widely spread all over the world to allow users to stay connected almost anywhere on the earth. When the signal (a beam of modulated microwave) is sent near the satellite then, satellite amplifies the signal and sent it back to the antenna receiver which is located on the surface of the earth. Satellite communication contains two main components like the space segment and the ground segment.The ground segment consists of fixed or mobile transmission, reception and ancillary equipment and the space segment, which mainly is the satellite itself.

Satellite communciaiton.

Infrared Communication

Infrared wireless communication communicates information in a device or systems through IR radiation. IR is electromagnetic energy at a wavelength that is longer than that of red light. It is used for security control, TV remote control and short range communications. In the electromagnetic spectrum, IR radiation lies between microwaves and visible light. So, they can be used as a source of communication.

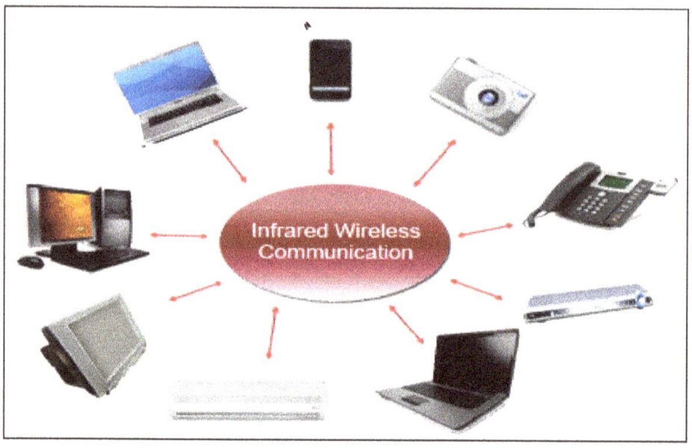

Infrared communication.

For a successful infrared communication, a photo LED transmitter and a photo diode receptor are required. The LED transmitter transmits the IR signal in the form of non visible light, that is captured and saved by the photoreceptor. So the information between the source and the target is transferred in this way. The source and destination can be mobile phones, TVs, security systems, laptops etc supports wireless communication.

Broadcast Radio

The first wireless communication technology is the open radio communication to seek out widespread use, and it still serves a purpose nowadays. Handy multichannel radios permit a user to speak over short distances, whereas citizen's band and maritime radios offer communication services for sailors. Ham radio enthusiasts share data and function emergency communication aids throughout disasters with their powerful broadcasting gear, and can even communicate digital information over the radio frequency spectrum.

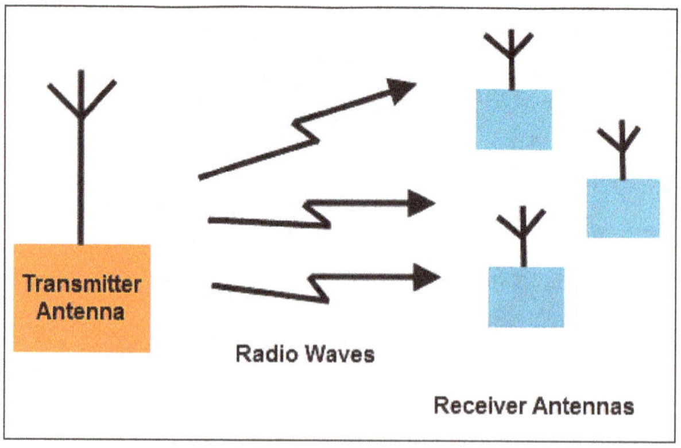

Broadcast radio.

Mostly an audio broadcasting service, radio broadcasts sound through the air as radio waves. Radio uses a transmitter which is used to transmit the data in the form of radio waves to a receiving antenna. To broadcast common programming, stations are associated with the radio N/W's. The broadcast happens either in simulcast or syndication or both. Radio broadcasting may be done via cable FM, the net and satellites. A broadcast sends information over long distances at up to two megabits/Sec (AM/FM Radio).

Radio waves are electromagnetic signals, that are transmitted by an antenna. These waves have completely different frequency segments, and you will be ready to obtain an audio signal by changing into a frequency segment.

Radio.

For example, you can take a radio station. When the RJ says you are listening to 92.7 BIG FM, what he really means is that signals are being broadcasted at a frequency of 92.7megahertz, that successively means the transmitter at the station is periodic at a frequency of 92.700,000 Cycles/second.

When you would like to listen to 92.7 BIG FM, all you have to do is tune the radio to just accept that specific frequency and you will receive perfect audio reception.

Microwave Communication

Microwave wireless communication is an effective type of communication, mainly this transmission uses radio waves, and the wavelengths of radio waves are measured in centimeters. In this communication, the data or information can be transfers using two methods. One is satellite method and another one is terrestrial method.

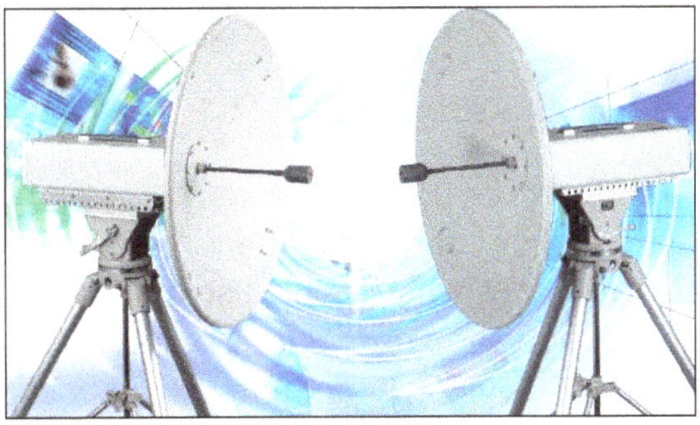

Microwave communication.

Wherein satellite method, the data can be transmitted through a satellite, that orbit 22,300 miles above the earth. Stations on the earth send and receive data signals from the satellite with a frequency ranging from 11GHz-14GHz and with a transmission speed of 1Mbps to 10Mbps. In terrestrial method, in which two microwave towers with a clear line of sight between them are used, ensuring no obstacles to disrupt the line of sight. So it is used often for the purpose of privacy. The frequency range of the terrestrial system is typically 4GHz-6GHz and with a transmission speed is usually 1Mbps to 10Mbps.

The main disadvantage of microwave signals is, they can be affected by bad weather, especially rain.

Wi-Fi

Wi-Fi is a low power wireless communication, that is used by various electronic devices like smart phones, laptops, etc.In this setup, a router works as a communication hub wirelessly. These networks allow users to connect only within close proximity to a router. WiFi is very common in networking applications which affords portability wirelessly. These networks need to be protected with passwords for the purpose of security, otherwise it will access by others.

Wi-Fi communication.

Mobile Communication Systems

The advancement of mobile networks is enumerated by generations. Many users communicate across a single frequency band through mobile phones. Cellular and cordless phones are two examples of devices which make use of wireless signals. Typically, cell phones have a larger range of networks to provide coverage. But, Cordless phones have a limited range. Similar to GPS devices, some phones make use of signals from satellites to communicate.

Mobile communication systems.

Bluetooth Technology

The main function of the Bluetooth technology is that permits you to connect a various electronic devices wirelessly to a system for the transferring of data.Cell phones are connected to hands free earphones, mouse, wireless keyboard. By using Bluetooth device the information from one device to another device. This technology has various functions and it is used commonly in the wireless communication market.

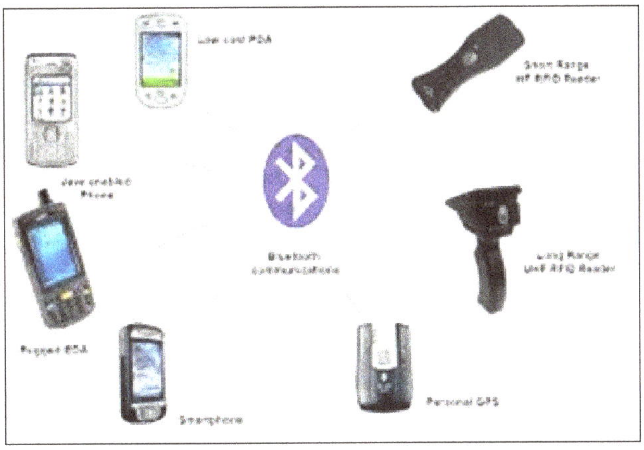

Bluetooth technology.

Antennas

Antenna are also called Aerial, component of radio, television, and radar systems that directs incoming and outgoing radio waves. Antennas are usually metal and have a wide variety of configurations, from the mastlike devices employed for radio and television broadcasting to the large parabolic reflectors used to receive satellite signals and the radio waves generated by distant astronomical objects.

The first antenna was devised by the German physicist Heinrich Hertz. During the late 1880s he carried out a landmark experiment to test the theory of the British mathematician-physicist James Clerk Maxwell that visible light is only one example of a larger class of electromagnetic effects that could pass through air (or empty space) as a succession of waves. Hertz built a transmitter for such waves consisting of two flat, square metallic plates, each attached to a rod, with the rods in turn connected to metal spheres spaced close together. An induction coil connected to the spheres caused a spark to jump across the gap, producing oscillating currents in the rods. The reception of waves at a distant point was indicated by a spark jumping across a gap in a loop of wire.

The Italian physicist Guglielmo Marconi, the principal inventor of wireless telegraphy, constructed various antennas for both sending and receiving, and he also discovered the importance of tall antenna structures in transmitting low-frequency signals. In the early antennas built by Marconi and others, operating frequencies were generally determined by antenna size and shape. In later antennas frequency was regulated by an oscillator, which generated the transmitted signal.

More powerful antennas were constructed during the 1920s by combining a number of elements in a systematic array. Metal horn antennas were devised during the subsequent decade following the development of waveguides that could direct the propagation of very high-frequency radio signals.

Over the years, many types of antennas have been developed for different purposes. An antenna may be designed specifically to transmit or to receive, although these functions may be performed by the same antenna. A transmitting antenna, in general, must be able to handle much more electrical energy than a receiving antenna. An antenna also may be designed to transmit at specific frequencies. In the United States, amplitude modulation (AM) radio broadcasting, for instance, is done at frequencies between 535 and 1,605 kilohertz (kHz); at these frequencies, a wavelength is hundreds of metres or yards long, and the size of the antenna is therefore not critical. Frequency modulation (FM) broadcasting, on the other hand, is carried out at a range from 88 to 108 megahertz (MHz). At these frequencies a typical wavelength is about 3 metres (10 feet) long, and the antenna must be adjusted more precisely to the electromagnetic wave, both in transmitting and in receiving. Antennas may consist of single lengths of wire or rods in various shapes (dipole, loop, and helical antennas), or of more elaborate arrangements of elements (linear, planar, or electronically steerable arrays). Reflectors and lens antennas use a parabolic dish to collect and focus the energy of radio waves, in much the same way that a parabolic mirror in a reflecting telescope collects light rays. Directional antennas are designed to be aimed directly at the signal source and are used in direction-finding.

Radar

Radar is an electromagnetic sensor used for detecting, locating, tracking, and recognizing objects of various kinds at considerable distances. It operates by transmitting electromagnetic energy toward objects, commonly referred to as targets, and observing the echoes returned from them. The targets may be aircraft, ships, spacecraft, automotive vehicles, and astronomical bodies, or even birds, insects, and rain. Besides determining the presence, location, and velocity of such objects, radar can sometimes obtain their size and shape as well. What distinguishes radar from optical and infrared sensing devices is its ability to detect faraway objects under adverse weather conditions and to determine their range, or distance, with precision.

Radar is an "active" sensing device in that it has its own source of illumination (a transmitter) for locating targets. It typically operates in the microwave region of the electromagnetic spectrum—measured in hertz (cycles per second), at frequencies extending from about 400 megahertz (MHz) to 40 gigahertz (GHz). It has, however, been used at lower frequencies for long-range applications (frequencies as low as several megahertz, which is the HF [high-frequency], or shortwave, band) and at optical and infrared frequencies (those of laser radar, or lidar). The circuit components and other hardware of radar systems vary with the frequency used, and systems range in size from those small enough to fit in the palm of the hand to those so enormous that they would fill several football fields.

Radar underwent rapid development during the 1930s and 1940s to meet the needs of the military. It is still widely employed by the armed forces, where many technological advances have originated. At the same time, radar has found an increasing number of important civilian applications, notably air traffic control, weather observation, remote sensing of the environment, aircraft and ship navigation, speed measurement for industrial applications and for law enforcement, space surveillance, and planetary observation.

Fundamentals of Radar

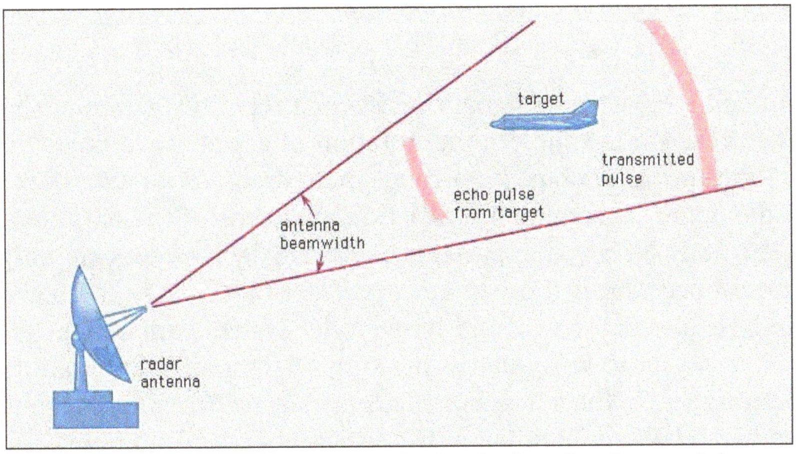

Principle of radar operation. The transmitted pulse has already passed the target, which has reflected a portion of the radiated energy back toward the radar unit.

Radar typically involves the radiating of a narrow beam of electromagnetic energy into space from an antenna. The narrow antenna beam scans a region where targets are expected. When a target is illuminated by the beam, it intercepts some of the radiated energy and reflects a portion back toward the radar system. Since most radar systems do not transmit and receive at the same time, a single antenna is often used on a time-shared basis for both transmitting and receiving.

A receiver attached to the output element of the antenna extracts the desired reflected signals and (ideally) rejects those that are of no interest. For example, a signal of interest might be the echo from an aircraft. Signals that are not of interest might be echoes from the ground or rain, which can mask and interfere with the detection of the desired echo from the aircraft. The radar measures the location of the target in range and angular direction. Range, or distance, is determined by measuring the total time it takes for the radar signal to make the round trip to the target and back. The angular direction of a target is found from the direction in which the antenna points at the time the echo signal is received. Through measurement of the location of a target at successive instants of time, the target's recent track can be determined. Once this information has been established, the target's future path can be predicted. In many surveillance radar applications, the target is not considered to be "detected" until its track has been established.

Pulse Radar

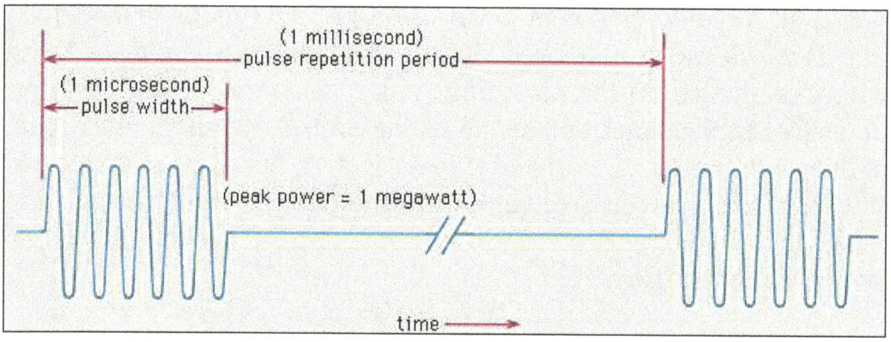

A typical pulse waveform transmitted by radar.

The most common type of radar signal consists of a repetitive strain of short-duration pulses. The figure shows a simple representation of a sine-wave pulse that might be generated by the transmitter of medium-range radar designed for aircraft detection. The sine wave in the figure represents the variation with time of the output voltage of the transmitter. The numbers given in parentheses in the figure are meant only to be illustrative and are not necessarily those of any particular radar. They are, however, similar to what might be expected for a ground-based radar system with a range of about 50 to 60 nautical miles (90 to 110 km), such as the kind used for air traffic control at airports. The pulse width is given in the figure as 1 microsecond (10−6 second). It should be noted that the pulse is shown as containing only a few cycles of the sine wave; however, in a radar system having the values indicated, there would be 1,000 cycles within the pulse. In the figure the time between successive pulses is given as 1 millisecond (10−3 second), which corresponds to a pulse repetition frequency of 1 kilohertz (kHz). The power of the pulse, called the peak power, is taken here to be 1 megawatt. Since pulse radar does not radiate continually, the average power is much less than the peak power. In this example, the average power is 1 kilowatt. The average power, rather than the peak power, is the measure of the capability of a radar system. Radars have average powers from a few milliwatts to as much as one or more megawatts, depending on the application.

A weak echo signal from a target might be as low as 1 picowatt (10–12 watt). In short, the power levels in a radar system can be very large (at the transmitter) and very small (at the receiver).

Another example of the extremes encountered in a radar system is the timing. Air-surveillance radar (one that is used to search for aircraft) might scan its antenna 360 degrees in azimuth in a few seconds, but the pulse width might be about one microsecond in duration. Some radar pulse widths are even of nanosecond (10^{-9} second) duration.

Radar waves travel through the atmosphere at roughly 300,000 km per second (the speed of light). The range to a target is determined by measuring the time that a radar signal takes to travel out to the target and back. The range to the target is equal to cT/2, where c = velocity of propagation of radar energy, and T = round-trip time as measured by the radar. From this expression, the round-trip travel of the radar signal through air is at a rate of 150,000 km per second. For example, if the time that it takes the signal to travel out to the target and back was measured by the radar to be 0.0006 second (600 microseconds), and then the range of the target would be 90 km. The ability to measure the range to a target accurately at long distances and under adverse weather conditions is radar's most distinctive attribute. There are no other devices that can compete with radar in the measurement of range.

The range accuracy of simple pulse radar depends on the width of the pulse: the shorter the pulse, the better the accuracy. Short pulses, however, require wide bandwidths in the receiver and transmitter (since bandwidth is equal to the reciprocal of the pulse width). Radar with a pulse width of one microsecond can measure the range to an accuracy of a few tens of metres or better. Some special radar can measure to an accuracy of a few centimetres. The ultimate range accuracy of the best radars is limited by the known accuracy of the velocity at which electromagnetic waves travel.

Directive Antennas and Target Direction

Almost all radars use a directive antenna–i.e., one that directs its energy in a narrow beam. (The beamwidth of an antenna of fixed size is inversely proportional to the radar frequency). The direction of a target can be found from the direction in which the antenna is pointing when the received echo is at a maximum. A precise means for determining the direction of a target is the monopulse method–in which information about the angle of a target is obtained by comparing the amplitudes of signals received from two or more simultaneous receiving beams, each slightly offset (squinted) from the antenna's central axis. A dedicated tracking radar—one that follows automatically a single target so as to determine its trajectory—generally has a narrow, symmetrical "pencil" beam. (A typical beamwidth might be about 1 degree). Such a radar system can determine the location of the target in both azimuth angle and elevation angle. Aircraft-surveillance radar generally employs an antenna that radiates a "fan" beam, one that is narrow in azimuth (about 1 or 2 degrees) and broad in elevation (elevation beamwidths of from 20 to 40 degrees or more). A fan beam allows only the measurement of the azimuth angle.

Doppler Frequency and Target Velocity

Radar can extract the Doppler frequency shift of the echo produced by a moving target by noting how much the frequency of the received signal differs from the frequency of the signal that was transmitted. (The Doppler Effect in radar is similar to the change in audible pitch experienced when a train whistle or the siren of an emergency vehicle moves past the listener). A moving target will cause the frequency of the echo signal to increase if it is approaching the radar or to decrease if it is receding from the radar. For example, if a radar system operates at a frequency of 3,000 MHz and an aircraft is moving toward it at a speed of 400 knots (740 km per hour), the frequency of the received echo signal will be greater than that of the transmitted signal by about 4.1 kHz. The Doppler frequency shift in hertz is equal to 3.4 $f_o v_r$, where fo is the radar frequency in gigahertz and v_r is the radial velocity (the rate of change of range) in knots.

Since the Doppler frequency shift is proportional to radial velocity, a radar system that measures such a shift in frequency can provide the radial velocity of a target. The Doppler frequency shift can also be used to separate moving targets from stationary targets even when the echo signal from undesired clutter is much more powerful than the echo from the desired moving targets. A form of pulse radar that uses the Doppler frequency shift to eliminate stationary clutter is called either moving-target indication (MTI) radar or a pulse Doppler radar, depending on the particular parameters of the signal waveform.

The above measurements of range, angle, and radial velocity assume that the target is a "point-scattered." Actual targets, however, are of finite size and can have distinctive shapes. The range profile of a finite-sized target can be determined if the range resolution of the radar is small compared with the target's size in the range dimension. (The range resolution of a radar, given in units of distance, is a measure of the ability of a radar to separate two closely spaced echoes). Some radars can have resolutions much smaller than one metre, which is quite suitable for determining the radial size and profile of many targets of interest.

The resolution in angle, or cross range, that can be obtained with conventional antennas is poor compared with that which can be obtained in range. It is possible, however, to achieve good resolution in angle by resolving in Doppler frequency (i.e., separating one Doppler frequency from another). If the radar is moving relative to the target (as when the radar is on an aircraft and the target is the ground), the Doppler frequency shift will be different for different parts of the target. Thus, the Doppler frequency shift can allow the various parts of the target to be resolved. The resolution in cross range derived from the Doppler frequency shift is far better than that achieved with a narrow-beam antenna. It is not unusual for the cross-range resolution obtained from Doppler frequency to be comparable to that obtained in the range dimension.

Radar Imaging

Radar can distinguish one kind of target from another (such as a bird from an aircraft), and some systems are able to recognize specific classes of targets (for example, a commercial airliner as opposed to a military jet fighter). Target recognition is accomplished by measuring the size and speed of the target and by observing the target with high resolution in one or more dimensions. Propellers and jet engines modify the radar echo from aircraft and can assist in target recognition. The flapping of the wings of a bird in flight produces a characteristic modulation that can be used to recognize that a bird is present or even to distinguish one type of bird from another.

Cross-range resolution obtained from Doppler frequency, along with range resolution, is the basis for synthetic aperture radar (SAR). SAR produces an image of a scene that is similar, but not identical, to an optical photograph. One should not expect the image seen by radar "eyes" to be the same as that observed by optical eyes. Each provides different information. Radar and optical images differ because of the large difference in the frequencies involved; optical frequencies are approximately 100,000 times higher than radar frequencies.

SAR can operate from long range and through clouds or other atmospheric effects that limit optical and infrared imaging sensors. The resolution of a SAR image can be made independent of range, an advantage over passive optical imaging where the resolution worsens with increasing range. Synthetic aperture radars that map areas of the Earth's surface with resolutions of a few metres can provide information about the nature of the terrain and what is on the surface.

A SAR operates on a moving vehicle, such as an aircraft or spacecraft, to image stationary objects or planetary surfaces. Since relative motion is the basis for the Doppler resolution, high resolution (in cross range) also can be accomplished if the radar is stationary and the target is moving. This is called inverse synthetic aperture radar (ISAR). Both the target and the radar can be in motion with ISAR.

Basic Radar System

The figure shows the basic parts of a typical radar system. The transmitter generates the high-power signal that is radiated by the antenna. In a sense, an antenna acts as a "transducer" to couple electromagnetic energy from the transmission line to radiation in space, and vice versa. The duplexer permits alternate transmission and reception with the same antenna; in effect, it is a fast-acting switch that protects the sensitive receiver from the high power of the transmitter.

The receiver selects and amplifies radar echoes so that they can be displayed on a television-like screen for the human operator or be processed by a computer. The signal processor separates the signals reflected by possible targets from unwanted clutter. Then, on the basis of the echo's exceeding a predetermined value, a human operator or a digital computer circuit decides whether a target is present.

Once it has been decided that a target is present and its location (in range and angle) has been determined, the track of the target can be obtained by measuring the target location at different times. During the early days of radar, target tracking was performed by an operator marking the location of the target "blip" on the face of a cathode-ray tube (CRT) display with a grease pencil. Manual tracking has been largely replaced by automatic electronic tracking, which can process hundreds or even thousands of target tracks simultaneously.

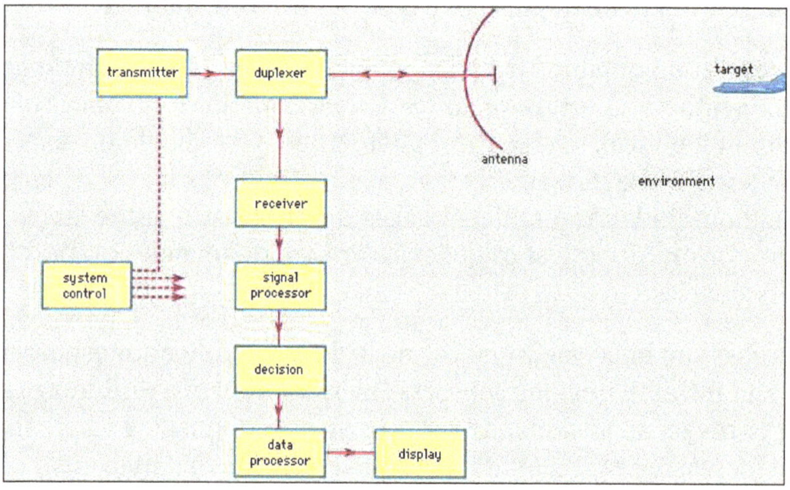

Basic parts of a radar system.

The system control optimizes various parameters on the basis of environmental conditions and provides the timing and reference signals needed to permit the various parts of the radar to operate effectively as an integrated system.

A widely used form of radar antenna is the parabolic reflector, the principle of which is shown in cross section in part A of the figure. A horn antenna or other small antenna is placed at the focus of the parabola to illuminate the parabolic surface of the reflector. After being reflected by this surface, the electromagnetic energy is radiated as a narrow beam. A paraboloid, which is generated by rotating a parabola about its axis, forms a symmetrical beam called a pencil beam. A fan beam, one with a narrow beamwidth in azimuth and a broad beamwidth in elevation, can be obtained by illuminating an asymmetrical section of the paraboloid. An example of an antenna that produces a fan beam is shown in the photograph.

The half-wave dipole (see part B of the figure), whose dimension is one-half of the radar wavelength, is the classic type of electromagnetic antenna. A single dipole is not of much use for radar, since it produces a beamwidth too wide for most applications. Radar requires a narrow beam (a beamwidth of only a few degrees) in order to concentrate its energy on the target and to determine the target location with accuracy. Such narrow beams can be formed by combining many individual dipole antennas so that the signals radiated or received by each elemental dipole are in unison, or in step. (The radar engineer would say that the signals are "in phase" with one another or that they are coherently added together). This is called a phased-array antenna, the basic principle of which is shown in part C of the figure.

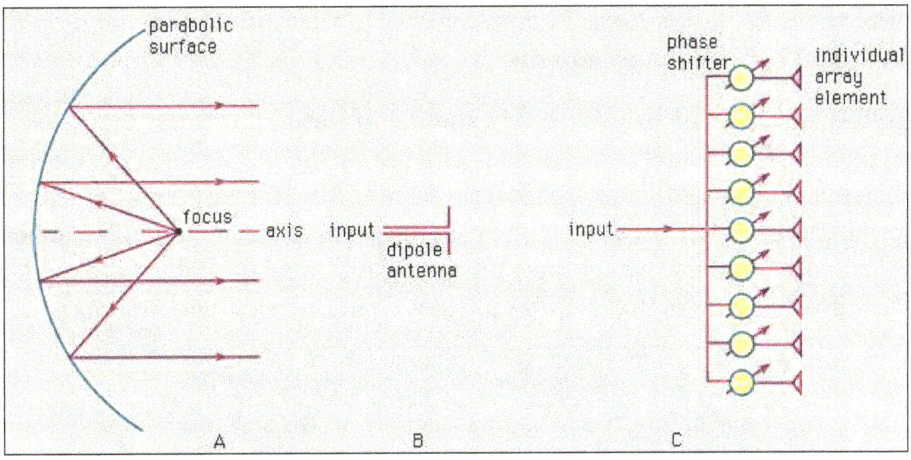

(A) A parabolic reflector antenna in which the energy radiated from the focus is reflected from the parabolic surface as a narrow beam. (B) A dipole antenna. (C) A phased-array antenna composed of many individual radiating elements.

The phase shifters at each radiating antenna-element change (or shift) the phase of the signal, so that all signals received from a particular direction will be in step with one another. As a result, the signals received at the elements add together without theoretical loss. Similarly, all signals radiated by the individual elements of the antenna will be in step with one another in some specific direction. Changing the phase shift at each element alters the direction of the antenna beam. An antenna of this kind is called an electronically steered phased-array. It allows rapid changes in the position of the beam without moving large mechanical structures. In some systems the beam can be changed from one direction to another within microseconds.

The individual radiating elements of a phased-array antenna need not be dipoles; various other types of antenna elements also can be used. For example, slots cut in the side of a waveguide are common, especially at the higher microwave frequencies. In a radar that requires a one-degree pencil-beam antenna, there might be about 5,000 individual radiating elements (the actual number depends on the particular design). The phased-array radar is more complex than radar systems that employ reflector antennas, but it provides capabilities not otherwise available.

Since there are many control points (each individual antenna element) in a phased-array, the radiated beam can be shaped to give a desired pattern to the beam. Controlling the shape of the radiated beam is important when the beam has to illuminate the airspace where aircraft are found but not illuminate the ground, where clutter echoes are produced. Another example is when the stray radiation (called antenna sidelobes) outside the main beam of the antenna pattern must be minimized.

The electronically steered phased-array is attractive for applications that require large antennas or when the beam must be rapidly changed from one direction to another. Satellite surveillance radars and long-range ballistic-missile-detection radars are examples that usually require phased-arrays. The U.S. Army's Patriot battlefield air-defense system and the U.S. Navy's Aegis system for ship air defense also depend on the electronically steered phased-array antenna.

The phased-array antenna is also used in some applications without the phase shifters shown in part C of the figure. The beam is steered by the mechanical movement of the entire antenna. Antennas of this sort are preferred over the parabolic reflector for airborne applications, in land-based air-surveillance radars requiring multiple beams (as in the so-called 3D radars, which measure elevation angle in addition to azimuth and range), and in applications that require ultralow antenna sidelobe radiation.

The AN/APG-66 radar in an F-16 fighter aircraft. The mechanically scanned planar phased-array antenna with radiating horizontal slots is 29 inches (74 cm) wide by 19 inches (48 cm) high.

Transmitter

The transmitter of a radar system must be efficient, reliable, not too large in size and weight, and easily maintained, as well as have the wide bandwidth and high power that are characteristic of radar applications. In general, the transmitter must generate low-noise, stable transmissions so that extraneous (unwanted) signals from the transmitter do not interfere with the detection of the small Doppler frequency shift produced by weak moving targets.

It is observed in the section History of radar that the invention of the magnetron transmitter in the late 1930s resulted in radar systems that could operate at the higher frequencies known as microwaves. The magnetron transmitter has certain limitations, but it continues to be used, for example, in low-average-power applications such as ship navigation radar and airborne weather-avoidance radar. The magnetron is a power oscillator in that it self-oscillates (i.e., generates microwave energy) when voltage is applied. Other radar transmitters usually are power amplifiers in that they take low-power signals at the input and amplify them to high power at the output. This provides stable high-power signals, as the signals to be radiated can be generated with precision at low power.

The klystron amplifier is capable of some of the highest power levels used in radar (many hundreds of kilowatts of average power). It has good efficiency and good stability. The disadvantages of the klystron are that it is usually large and it requires high voltages (e.g., about 90 kilovolts for one megawatt of peak power). At low power the instantaneous bandwidth of the klystron is small, but the klystron is capable of large bandwidth at high peak powers of a few megawatts.

The traveling-wave tube (TWT) is related to the klystron. It has very wide bandwidths at low peak power, but, as the peak power levels are increased to those needed for pulse radar, its bandwidth decreases. As peak power increases, the bandwidths of the TWT and the klystron approach one another.

Solid-state transmitters, such as the transistor, are attractive because of their potential for long life, ease of maintenance, and relatively wide bandwidth. An individual solid-state device generates relatively low power and can be used only when the radar application can be accomplished with low power (as in short-range applications or in the radar altimeter). High power can be achieved, however, by combining the outputs of many individual solid-state devices.

While the solid-state transmitter is easy to maintain and is capable of wide-band operation, it has certain disadvantages. It is much better suited for long pulses (milliseconds) than for short pulses (microseconds). Long pulses can complicate radar operation because signal processing (such as pulse compression) is needed to achieve the desired range resolution. Furthermore, a long-pulse radar generally requires several different pulse widths: a long pulse for long range and one or more shorter, high-energy pulses with less energy to observe targets at the ranges masked when the long pulse is transmitting. (A one-millisecond pulse, for example, masks echoes from 0 to about 80 nautical miles, or 150 km).

Every kind of transmitter has its disadvantages as well as advantages. In any particular application, the radar engineer must continually search for compromises that give the results desired without too many negative effects that cannot be adequately accommodated.

Receivers

Like most other receivers, the radar receiver is a classic super heterodyne. It has to filter the desired echo signals from clutter and receiver noise that interfere with detection. It also must amplify the weak received signals to a level where the receiver output is large enough to actuate a display or a computer. The technology of the radar receiver is well established and seldom sets a limit on radar performance.

The receiver must have a large dynamic range in situations where it is necessary to detect weak signals in the presence of very large clutter echoes by recognizing the Doppler frequency shift of the desired moving targets. Dynamic range can be loosely described as the ratio of the strongest to the weakest signals that can be handled without significant distortion by a receiver. A radar receiver might be required to detect signals that vary in power by a million to one—and sometimes much more.

In most cases the sensitivity of a radar receiver is determined by the noise generated internally at its input. Because it does not generate much noise of its own, a transistor is usually used as the first stage of a receiver.

Signal and Data Processors

The signal processor is the part of the receiver that extracts the desired target signal from unwanted clutter. It is not unusual for these undesired reflections to be much larger than desired target echoes, in some cases more than one million times larger. Large clutter echoes from stationary objects can be separated from small moving target echoes by noting the Doppler frequency shift produced by the moving targets. Most signal processing is performed digitally with computer technology. Digital processing has significant capabilities in signal processing not previously available with analog methods.

Pulse compression is sometimes included under signal processing. It too benefits from digital technology, but analog processors (e.g., surface acoustic wave delay lines) are used rather than digital methods when pulse compression must achieve resolutions of a few feet or less.

Displays

Although it has its limitations, the cathode-ray tube (CRT) has been the preferred technology for displaying information ever since the early days of radar. There have been, however, considerable improvements in flat-panel displays because of the demands of computers and television. Flat-panel displays occupy less volume and require less power than CRTs, but they also have their limitations. Radar has taken advantage of flat-panel displays and has become increasingly important as a display.

In the early days of radar, an operator decided whether a target was present on the basis of what raw data were displayed. Modern radars, however, present processed information to the operator. Detections are made automatically in the receiver without operator involvement and are then presented on the display to the operator for further action.

A commonly used radar display is the plan position indicator (PPI), which provides a maplike presentation in polar coordinates of range and angle. The display is "dark" except when echo signals are present.

All practical radar displays have been two-dimensional, yet many types of radar provide more information than can be displayed on the two coordinates of a flat screen. Colour coding of the signal indicated on the PPI is sometimes used to provide additional information about the echo signal. Colour has been employed, for example, to indicate the strength of the echo. Doppler weather radars make good use of colour coding to indicate on a two-dimensional display the levels of rain intensity associated with each echo shown. They also utilize colour to indicate the radial speed of the wind, the wind shear, and other information relating to severe storms.

Factors affecting Radar Performance

The performance of a radar system can be judged by the following: (1) the maximum range at which it can see a target of a specified size, (2) the accuracy of its measurement of target location in range and angle, (3) its ability to distinguish one target from another, (4) its ability to detect the desired target echo when masked by large clutter echoes, unintentional interfering signals from other "friendly" transmitters, or intentional radiation from hostile jamming (if a military radar), (5) its ability to recognize the type of target, and (6) its availability (ability to operate when needed), reliability, and maintainability.

Transmitter Power and Antenna Size

The maximum range of a radar system depends in large part on the average power of its transmitter and the physical size of its antenna. (In technical terms, this is called the power-aperture product). There are practical limits to each. As noted before, some radar systems have an average power of roughly one megawatt. Phased-array radars about 100 feet (30 metres) in diameter are not uncommon; some are much larger. There are specialized radars with (fixed) antennas, such as some HF over-the-horizon radars and the U.S. Space Surveillance System (SPASUR), that extend more than one mile (1.6 km).

Receiver Noise

The sensitivity of a radar receiver is determined by the unavoidable noise that appears at its input. At microwave radar frequencies, the noise that limits detectability is usually generated by the receiver itself (i.e., by the random motion of electrons at the input of the receiver) rather than by external noise that enters the receiver via the antenna. A radar engineer often employs a transistor amplifier as the first stage of the receiver even though lower noise can be obtained with more sophisticated (and more complex) devices. This is an example of the application of the basic engineering principle that the "best" performance that can be obtained might not necessarily be the solution that best meets the needs of the user.

The receiver is designed to enhance the desired signals and to reduce the noise and other undesired signals that interfere with detection. A designer attempts to maximize the detectability of weak signals by using what radar engineers call a "matched filter," which is a filter that maximizes the signal-to-noise ratio at the receiver output. The matched filter has a precise mathematical formulation that depends on the shape of the input signal and the character of the receiver noise. A suitable approximation to the matched filter for the ordinary pulse radar, however, is one whose bandwidth in hertz is the reciprocal of the pulse width in seconds.

Target Size

The size of a target as "seen" by radar is not always related to the physical size of the object. The measure of the target size as observed by radar is called the radar cross section and is given in units of area (square metres). It is possible for two targets with the same physical cross-sectional area to differ considerably in radar size, or radar cross section. For example, a flat plate 1 square metre in area will produce a radar cross section of about 1,000 square metres at a frequency of 3 GHz when viewed perpendicular to the surface. A cone-sphere (an object resembling an ice-cream cone) when viewed in the direction of the cone rather than the sphere could have a radar cross section of about 0.001 square metre even though its projected area is also 1 square metre. In theory, the radar cross section has little to do with the size of the cone or the cone angle. Thus, the flat plate and the cone-sphere can have radar cross sections that differ by a million to one even though their physical projected areas are the same.

The sphere is an unusual target in that its radar cross section is the same as its physical cross-sectional area (when its circumference is large compared with the radar wavelength). That is to say, a sphere with a projected area of 1 square metre has a radar cross section of 1 square metre.

Commercial aircraft might have radar cross sections from about 10 to 100 square metres, except when viewed broadside, where the cross sections are much larger. Most air-traffic-control radars are required to detect aircraft with a radar cross section as low as 2 square metres, since some small general-aviation aircraft can be of this value. For comparison, the radar cross section of a man has been measured at microwave frequencies to be about 1 square metre. A bird can have a cross section of 0.01 to 0.001 square metre. Although this is a small value, a bird can be readily detected at ranges of several tens of kilometres by long-range radar. In general, many birds can be detected by radar, so special measures must usually be taken to ensure that their echoes do not interfere with the detection of desired targets.

The radar cross section of an aircraft and that of most other targets of practical interest fluctuate rapidly as the aspect of the target changes with respect to the radar unit. It would not be unusual for a slight change in aspect to cause the radar cross section to change by a factor of 10 to 1,000.

Clutter

Echoes from land, sea, rain, snow, hail, birds, insects, auroras, and meteors are of interest to those who observe and study the environment, but they are a nuisance to those who want to detect aircraft, ships, missiles, or other similar targets. Clutter echoes can seriously limit the capability of a radar system; thus, a significant part of radar design is devoted to minimizing the effects of clutter without reducing the echoes from desired targets. The Doppler frequency shift is the usual means by which moving targets are distinguished from the clutter of stationary objects. Detection of targets in rain is less of a problem at the lower frequencies, since the radar echo from rain decreases rapidly with decreasing frequency and the average cross section of aircraft is relatively independent of frequency in the microwave region. Because raindrops are more or less spherical (symmetrical) and aircraft are asymmetrical, the use of circular polarization can enhance the detection of aircraft in rain. With circular polarization, the electric field rotates at the radar frequency. Because of this, the electromagnetic energy reflected by the rain and the aircraft will be affected differently, which thereby makes it easier to distinguish between the two. (In fair weather most radars use linear polarization; i.e., the direction of the electric field is fixed).

Atmospheric Effects

Rain and other forms of precipitation can cause echo signals that mask the desired target echoes. There are other atmospheric phenomena that can affect radar performance as well. The decrease in density of the Earth's atmosphere with increasing altitude causes radar waves to bend as they propagate through the atmosphere. This usually increases the detection range at low angles to a slight extent. The atmosphere can form "ducts" that trap and guide radar energy around the curvature of the Earth and allow detection at ranges beyond the normal horizon. Ducting over water is more likely to occur in tropical climates than in colder regions. Ducts can sometimes extend the range of airborne radar, but on other occasions they may cause the radar energy to be diverted and not illuminate regions below the ducts. This results in the formation of what are called radar holes in the coverage. Since it is not predictable or reliable, ducting can in some instances be more of a nuisance than a help.

Loss of radar energy due to atmospheric absorption, when propagation is through the clear atmosphere or rain, is usually small for most systems operating at microwave frequencies.

Interference

Signals from nearby radars and other transmitters can be strong enough to enter a radar receiver and produce spurious responses. Well-trained operators are not often deceived by interference, though they may find it a nuisance. Interference is not as easily ignored by automatic detection and tracking systems, however, and so some method is usually needed to recognize and remove interference pulses before they enter the automatic detector and tracker of radar.

Electronic Countermeasures (Electronic Warfare)

The purpose of hostile electronic countermeasures (ECM) is to degrade the effectiveness of military radar deliberately. ECM can consist of (1) noise jamming that enters the receiver via the antenna and increases the noise level at the input of the receiver, (2) false target generation, or repeater jamming, by which hostile jammers introduce additional signals into the radar receiver in an attempt to confuse the receiver into thinking that they are real target echoes, (3) chaff, which is an artificial cloud consisting of a large number of tiny metallic reflecting strips that create strong echoes over a large area to mask the presence of real target echoes or to create confusion, and (4) decoys, which are small, inexpensive air vehicles or other objects designed to appear to the radar as if they are real targets. Military radars are also subject to direct attack by conventional weapons or by antiradiation missiles (ARMs) that use radar transmissions to find the target and home in on it. A measure of the effectiveness of military radar is the large sums of money spent on electronic warfare measures, ARMs, and low-cross-section (stealth) aircraft.

Military radar engineers have developed various ways of countering hostile ECM and maintaining the ability of a radar system to perform its mission. It might be noted that a military radar system can often accomplish its mission satisfactorily even though its performance in the presence of ECM is not what it would be if such measures were absent.

Examples of Radar Systems

Airport Surveillance Radar

Airport surveillance radar systems are capable of reliably detecting and tracking aircraft at altitudes below 25,000 feet (7,620 metres) and within 40 to 60 nautical miles (75 to 110 km) of their airport. Systems of this type have been installed at more than 100 major airports throughout the United States. One such system, the ASR-9, is designed to be operable at least 99.9 percent of the time, which means that the system is down less than 10 hours per year. This high availability is attributable to reliable electronic components, a "built-in test" to search for failures, remote monitoring, and redundancy (i.e., the system has two complete channels except for the antenna; when one channel must be shut down for repair, the other continues to operate). The ASR-9 is designed to operate unattended, with no maintenance personnel at the radar site. A number of radar units can be monitored and controlled from a single location. When trouble occurs, the fault is identified and a maintenance person dispatched for repair.

Echoes from rain that mask the detection of aircraft are reduced by the use of Doppler filtering and other techniques devised to separate moving aircraft from undesired clutter. It is important, however, for air traffic controllers to recognize areas of severe weather so that they can direct aircraft safely around, rather than through, rough or hazardous conditions. The ASR-9 has a separate receiving channel that recognizes weather echoes and provides their location to air traffic controllers. Six different levels of precipitation intensity can be displayed, with or without the aircraft targets superimposed.

The ASR-9 system operates at frequencies from 2.7 to 2.9 GHz (within the S band). Its klystron transmitter has a peak power of 1.3 megawatts, a pulse width of 1 microsecond, and an antenna with a horizontal beamwidth of 1.4 degrees that rotates at 12.5 revolutions per minute (4.8-second rotation period).

It is 16.5 feet (5 metres) wide and 9 feet (2.75 metres) high. Atop the radar antenna (riding piggyback) is a lightweight planar-array antenna for the air-traffic-control radar-beacon system (ATCRBS). Its dimensions are 26 feet (8 metres) by 5.2 feet (1.6 metres). ATCRBS is the primary means for detecting and identifying aircraft equipped with a transponder that can reply to the ATCRBS interrogation. The

ATCRBS transmitter, which is independent of the radar system and operates at a different frequency, radiates a coded interrogation signal. Aircraft equipped with a suitable transponder can recognize the interrogation and send a coded reply at a frequency different from the interrogation frequency. The interrogator might then ask the aircraft, by means of other coded signals, to automatically identify itself and to report its altitude. ATCRBS works only with cooperative targets (i.e., those with an operational transponder).

The ASR-11 and ASR-12 are airport surveillance radars that utilize a solid-state (transistor) transmitter and long pulses rather than a klystron transmitter and short pulses.

Doppler Weather Radar

For many years radar has been used to provide information about the intensity and extent of rain and other forms of precipitation. This application of radar is well known in the United States from the familiar television weather reports of precipitation observed by the radars of the National Weather Service. A major improvement in the capability of weather radar came about when engineers developed new radars that could measure the Doppler frequency shift in addition to the magnitude of the echo signal reflected from precipitation. The Doppler frequency shift is important because it is related to the radial velocity of the precipitation blown by wind (the component of the wind moving either toward or away from the radar installation). Since tornadoes, mesocyclones (which spawn tornadoes), hurricanes, and other hazardous weather phenomena tend to rotate, measurement of the radial wind speed as a function of viewing angle will identify rotating weather patterns. (Rotation is indicated when the measurement of the Doppler frequency shift shows that the wind is coming toward the radar at one angle and away from it at a nearby angle).

The pulse Doppler weather radars employed by the National Weather Service, which are known as Nexrad, make quantitative measurements of precipitation, warn of potential flooding or dangerous hail, provide wind speed and direction, indicate the presence of wind shear and gust fronts, track storms, predict thunderstorms, and provide other meteorological information. In addition to measuring precipitation (from the intensity of the echo signal) and radial speed (from the Doppler frequency shift), Nexrad also measures the spread in radial speed (difference between the maximum and the minimum speeds) of the precipitation particles within each radar resolution cell. The spread in radial speed is an indication of wind turbulence.

Another improvement in the weather information provided by Nexrad is the digital processing of radar data, a procedure that renders the information in a form that can be interpreted by an observer who is not necessarily a meteorologist. The computer automatically identifies severe weather effects and indicates their nature on a display viewed by the observer. High-speed communication lines integrated with the Nexrad system allow timely weather information to be transmitted for display to various users.

The Nexrad radar operates at S-band frequencies (2.7 to 3 GHz) and is equipped with a 28-foot- (8.5-metre-) diameter antenna. It takes five minutes to scan its 1 degree beamwidth through 360 degrees in azimuth and from 0 to 20 degrees in elevation. The Nexrad system can measure rainfall up to a distance of 460 km and determine its radial velocity as far as 230 km.

A serious weather hazard to aircraft in the process of landing or taking off from an airport is the downburst, or microburst. This strong downdraft causes wind shear capable of forcing aircraft to the ground. Terminal Doppler weather radar (TDWR) is the name of the type of system at or near airports that is specially designed to detect dangerous microbursts. It is similar in principle to Nexrad but is a shorter-range system since it has to observe dangerous weather phenomena only in the vicinity of an airport. It operates from 5.60 to 5.65 GHz (C band) to avoid interference with the lower frequencies of Nexrad and ASR systems.

Airborne Combat Radar

A modern combat aircraft is generally required not only to intercept hostile aircraft but also to attack surface targets on the ground or sea. The radar that serves such an aircraft must have the capabilities to perform these distinct military missions. This is not easy because each mission has different requirements. The different ranges, accuracies, and rates at which the radar data is required, the effect of the environment (land or sea clutter), and the type of target (land features or moving aircraft) call for different kinds of radar waveforms (different pulse widths and pulse repetition frequencies). In addition, an appropriate form of signal processing is required to extract the particular information needed for each military function. Radar for combat aircraft must therefore be multimode—i.e., operate with different waveforms, signal processing, and antenna scanning. It would not be unusual for an airborne combat radar to have from 8 to 10 air-to-air modes and 6 to 10 air-to-surface modes. Furthermore, the radar system might be required to assist in rendezvous with a companion combat craft or with a refueling aircraft, provide guidance of air-to-air missiles, and counter hostile electronic jamming. The problem of achieving effectiveness with these many modes is a challenge for radar designers and is made more difficult by the size and weight constraints on combat aircraft.

The AN/APG-68(V)XM radar built for the U.S. F-16 (C/D) fighter. This is a pulse Doppler radar system that operates in a portion of the X-band (8- to 12-GHz) region of the spectrum. It occupies a volume of less than 0.13 cubic metre (4.6 cubic feet), weighs less than 164 kg (362 pounds), and requires an input power of 5.6 kilowatts. It can search 120 degrees in azimuth and elevation and is supposed to have a range of 35 nautical miles (65 km) in the "look-up" mode and 27.5 nautical miles (50 km) in the "look-down" mode. The look-up mode is a more or less conventional radar mode with a low pulse-repetition-frequency (prf) that is used when the target is at medium or high altitude and no ground-clutter echoes are present to mask target detection. The look-down mode uses a medium-prf Doppler waveform and signal processing that provide target detection in the presence of heavy clutter. (A low prf for an X-band combat radar might be from 250 hertz to 5 kHz, a medium prf from 5 to 20 kHz, and a high prf from 100 to 300 kHz.) Radars for larger combat aircraft can have greater capability but are, accordingly, bigger and heavier than the system just described.

The AN/APG-77 radar for the U.S. Air Force F-22 stealth dual-role fighter employs what is called an active-aperture phased-array radar rather than a mechanically scanned planar-array antenna. At each radiating element of the active-aperture phased-array is an individual transmitter, receiver, phase shifter, duplexer, and control.

Ballistic Missile Defense and Satellite-surveillance Radars

The systems for detecting and tracking ballistic missiles and orbiting satellites are much larger than those for aircraft detection because the ranges are longer and the radar echoes from space targets can be smaller than echoes from aircraft. Such radars might be required to have maximum ranges of 2,000 to 3,000 nautical miles (3,700 to 5,600 km), as compared with 200 nautical miles (370 km) for a typical long-range aircraft-detection system. The average power of the transmitter for a ballistic missile defense (BMD) radar can be from several hundred kilowatts to one megawatt or more, which is about 100 times greater than the average power of radars designed for aircraft detection. Antennas for this application have dimensions on the order of tens of metres to a hundred metres or more and are electronically scanned phased-array antennas capable of steering the radar beam without moving large mechanical structures. Radar systems for long-range ballistic missile detection and satellite surveillance are commonly found at the lower frequencies (typically at frequency bands of 420–450 MHz and 1,215–1,400 MHz).

The Pave Paws radar (AN/FPS-115) is an ultrahigh-frequency (UHF; 420–450 MHz) phased-array system for detecting submarine-launched ballistic missiles. It is supposed to detect targets with a radar cross section of 10 square metres at a range of 3,000 nautical miles (5,600 km). The array antenna contains 1,792 active elements within a diameter of 72.5 feet (22 metres). Each active element is a module with its own solid-state transmitter, receiver, duplexer, and phase shifter. The total average power per antenna is about 145 kilowatts. Two antennas make up a system, with each capable of covering a sector 120 degrees in azimuth. Vertical coverage is from 3 to 85 degrees. An upgraded variant of this type of radar is used in the Ballistic Missile Early Warning System (BMEWS) network, with installations in Alaska, Greenland, and England. BMEWS is designed to provide warning of intercontinental ballistic missiles (ICBMs). Each array antenna measures about 82 feet (25 metres) across and has 2,560 active elements identical to those of the Pave Paws system. Both the BMEWS and Pave Paws radars detect and track satellites and other space objects in addition to warning of the approach of ballistic missiles.

BMD radar has to engage one or more relatively small reentry vehicles (RVs) that carry a warhead. Ballistic missile RVs can be made to have a very low echo (low radar cross section) when illuminated by radar. They were the original low-radar-cross-section targets and appeared more than 20 years before the more highly publicized stealth aircraft became a reality in the late 1980s. Ballistic missile defense requires battle-management radars that not only detect and track a relatively small target at sufficient range to engage effectively but also must reliably distinguish the reentry vehicles that carry warheads from the many confusion targets that can be present. Confusion targets include decoys, chaff (strips of metallic foil that produce an echo similar in size to that of the reentry vehicle), exploded tank fragments, and other objects released by the attacking missile. The BMD radar must also be able to fulfill its mission in spite of hostile countermeasures and defend against ballistic missiles that can reenter at low angles (depressed trajectories). In addition, the radar must be located in a defended region and be hardened to survive either a conventional or a nuclear attack.

There are at least two basic approaches to ballistic missile defense depending on whether the RV is engaged outside the atmosphere (exoatmospheric) or within the atmosphere (endoatmospheric). Exoatmospheric engagement is attractive, since it occurs at long range and a single system can defend a large area, but it requires some reliable method to select the warhead from the many extraneous objects that can accompany the warhead. An endoatmospheric ballistic missile defense system takes advantage of the slowing down of the lighter objects (decoys, chaff, and fragments) when they reenter the atmosphere and encounter air resistance. After reentry, the heavy warhead will be separated from the accompanying lighter "junk" and thus can be engaged. A significant limitation, however, is that endoatmospheric ballistic missile defense results in a much smaller defended area.

Ground-probing Radar

Radar waves are usually thought of as being reflected from the surface of the ground. However, at the lower frequencies (below several hundred megahertz), radar energy can penetrate into the ground and be reflected from buried objects. The loss in propagating in the ground is very high at these frequencies, but it is low enough to permit ranges of about 3.3 to 33 feet (1 to 10 metres) or more. This is sufficient for probing the subsurface soil in order to detect underground tunnels and utility pipes and cables, to aid in archaeological digs, and to monitor the subsurface conditions of highways and bridge roadways. The short ranges require that the radar system be able to resolve closely spaced objects, which means wide-bandwidth signals must be radiated. Normally, wide bandwidth is not available at the lower frequencies (especially when a 1-foot (30-cm) range resolution requires a 500-MHz bandwidth). However, since the energy is directed into the ground rather than radiated into space, the large frequency band needed for high resolution can be obtained without serious interference to other users of the radio spectrum.

A ground-probing radar might radiate over frequencies ranging from 5 to 500 MHz in order to obtain good penetration (which requires low frequencies) with high resolution (which requires wide bandwidth). The antenna can be placed directly on the ground. Ground-probing radar units generally are small enough to be portable.

Over-the-horizon Radar

Frequencies lower than about 100 MHz usually are not desirable for radar applications. An example where lower frequencies can provide a unique and important capability is in the shortwave, or high-frequency (HF), portion of the radio band (from 3 to 30 MHz). The advantage of the HF band is that radio waves of these frequencies are refracted (bent) by the ionosphere so that the waves return to the Earth's surface at long distances beyond the horizon, as. This permits target detection at distances from about 500 to 2,000 nautical miles (900 to 3,700 km). Thus, an HF over-the-horizon (OTH) radar can detect aircraft at distances up to 10 times that of a ground-based microwave air-surveillance radar, whose range is limited by the curvature of the Earth. Besides detection and tracking of aircraft at long ranges, an HF OTH radar can be designed to detect ballistic missiles (particularly the disturbance caused by ballistic missiles as they travel through the ionosphere), ships, and weather effects over the ocean. Winds over the ocean generate waves on the water that can be recognized by HF OTH radar. From the Doppler frequency spectrum produced by echoes from the water waves, one can determine the direction of the waves generated by the wind and hence the direction of the wind itself. The strength of the waves (which indicates the state of the sea, or roughness) also can be ascertained. Timely information about the winds that drive waves over a wide expanse of the ocean can be obtained with HF OTH—obtainable only with great difficulty, if at all, by other means—which has proved valuable for weather prediction.

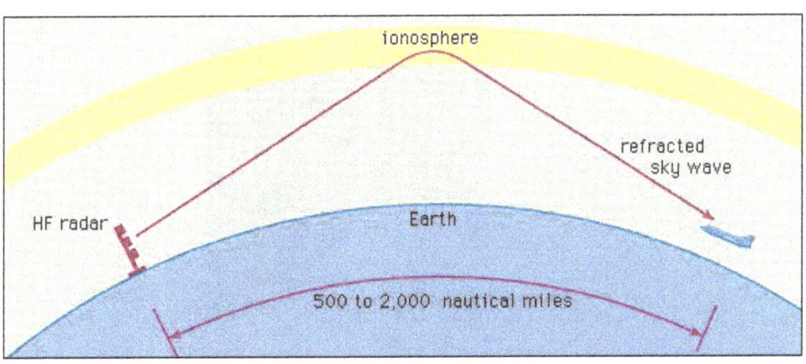

Refraction of HF radar radiation by the ionosphere.

An HF OTH radar might have an average power of about one megawatt and have phased-array antennas that sometimes extend several thousands of feet. This type of radar was originally developed for military purposes, especially for the surveillance of aircraft and ships over large expanses of water, where it is difficult for conventional microwave radars to provide coverage of large areas. For example, an important application of HF OTH is to provide wide-area surveillance of regions where illegal drug-carrying aircraft are suspected of operating. The area that can be held under surveillance by HF OTH radar is so large that it is difficult for aircraft to avoid detection by flying around or under its coverage. Furthermore, these counternarcotic radars can in many cases detect aircraft as they take off from a distant airfield and can sometimes follow them all the way to their destination. It is also possible in some cases to recognize specific aircraft types on the basis of the radar observation of the aircraft during take-off and landing. The U.S. Navy's HF OTH radars known as relocatable over-the-horizon radar (ROTHR), or AN/TPS-71, have been redirected for use in drug interdiction. Such radars, located in Virginia, Texas, and Puerto Rico, provide multiple coverage of drug-traffic regions in Central America and the northern part of South America. An ROTHR can cover a 64-degree wedge-shaped area at ranges from 500 to 1,600 nautical miles (900 to 3,000 km). Its receiving antenna is an electronically steered phased-array consisting of 372 pairs of monopole antennas. The antenna extends 1.4 nautical miles (2.5 km) in length. The transmitter operates from 5 to 28 MHz with an average power of 210 kilowatts. Each radar can provide surveillance of approximately 1.3 million square nautical miles (4.5 million square km). This is much more than 10 times the area covered by conventional surface-based long-range microwave air-surveillance radar.

Satellite Communication

Satellite communication in telecommunications is the use of artificial satellites to provide communication links between various points on Earth. Satellite communications play a vital role in the global telecommunications system. Approximately 2,000 artificial satellites orbiting Earth relay analog and digital signals carrying voice, video, and data to and from one or many locations worldwide.

AUSSAT-1 communications satellite being released in low Earth orbit. The satellite subsequently was boosted into a geostationary orbit by means of an attached rocket motor.

Satellite communication has two main components: the ground segment, which consists of fixed or mobile transmission, reception, and ancillary equipment, and the space segment, which primarily is the satellite itself. A typical satellite link involves the transmission or uplinking of a signal from an Earth station to a satellite. The satellite then receives and amplifies the signal and retransmits it back to Earth, where it is received and reamplified by Earth stations and terminals. Satellite receivers on the ground include direct-to-home (DTH) satellite equipment, mobile reception equipment in aircraft, satellite telephones, and handheld devices.

Working of a Satellite

A satellite is basically a self-contained communications system with the ability to receive signals from Earth and to retransmit those signals back with the use of a transponder—an integrated receiver and transmitter of radio signals. A satellite has to withstand the shock of being accelerated during launch up to the orbital velocity of 28,100 km (17,500 miles) an hour and a hostile space environment where it can be subject to radiation and extreme temperatures for its projected operational life, which can last up to 20 years. In addition, satellites have to be light, as the cost of launching a satellite is quite expensive and based on weight. To meet these challenges, satellites must be small and made of lightweight and durable materials. They must operate at a very high reliability of more than 99.9 percent in the vacuum of space with no prospect of maintenance or repair.

Intelsat VI, a communications satellite.

The main components of a satellite consist of the communications system, which includes the antennas and transponders that receive and retransmit signals, the power system, which includes the solar panels that provide power, and the propulsion system, which includes the rockets that propel the satellite. A satellite needs its own propulsion system to get itself to the right orbital location and to make occasional corrections to that position. A satellite in geostationary orbit can deviate up to a degree every year from north to south or east to west of its location because of the gravitational pull of the Moon and Sun. A satellite has thrusters that are fired occasionally to make adjustments in its position. The maintenance of a satellite's orbital position is called "station keeping," and the corrections made by using the satellite's thrusters are called "attitude control." A satellite's life span is determined by the amount of fuel it has to power these thrusters. Once the fuel runs out, the satellite eventually drifts into space and out of operation, becoming space debris.

A satellite in orbit has to operate continuously over its entire life span. It needs internal power to be able to operate its electronic systems and communications payload. The main source of power is sunlight, which is harnessed by the satellite's solar panels. A satellite also has batteries on board to provide power when the Sun is blocked by Earth. The batteries are recharged by the excess current generated by the solar panels when there is sunlight.

Satellites operate in extreme temperatures from −150 °C (−238 °F) to 150 °C (300 °F) and may be subject to radiation in space. Satellite components that can be exposed to radiation are shielded with aluminium and other radiation-resistant material. A satellite's thermal system protects its sensitive electronic and mechanical components and maintains it in its optimum functioning temperature to ensure its continuous operation. A satellite's thermal system also protects sensitive satellite components from the extreme changes in temperature by activation of cooling mechanisms when it gets too hot or heating systems when it gets too cold.

The tracking telemetry and control (TT&C) system of a satellite is a two-way communication link between the satellite and TT&C on the ground. This allows a ground station to track a satellite's position and control the satellite's propulsion, thermal, and other systems. It can also monitor the temperature, electrical voltages, and other important parameters of a satellite.

Communication satellites range from microsatellites weighing less than 1 kg (2.2 pounds) to large satellites weighing over 6,500 kg (14,000 pounds). Advances in miniaturization and digitalization have substantially increased the capacity of satellites over the years. Early Bird had just one transponder capable of sending just one TV channel. The Boeing 702 series of satellites, in contrast, can have more than 100 transponders, and with the use of digital compression technology each transponder can have up to 16 channels, providing more than 1,600 TV channels through one satellite.

Satellites operate in three different orbits: low Earth orbit (LEO), medium Earth orbit (MEO), and geostationary or geosynchronous orbit (GEO). LEO satellites are positioned at an altitude between 160 km and 1,600 km (100 and 1,000 miles) above Earth. MEO satellites operate from 10,000 to 20,000 km (6,300 to 12,500 miles) from Earth. (Satellites do not operate between LEO and MEO because of the inhospitable environment for electronic components in that area, which is caused by the Van Allen radiation belt.) GEO satellites are positioned 35,786 km (22,236 miles) above Earth, where they complete one orbit in 24 hours and thus remain fixed over one spot. It only takes three GEO satellites to provide global coverage, while it takes 20 or more satellites to cover the entire Earth from LEO and 10 or more in MEO. In addition, communicating with satellites in LEO and MEO requires tracking antennas on the ground to ensure seamless connection between satellites.

A signal that is bounced off a GEO satellite takes approximately 0.22 second to travel at the speed of light from Earth to the satellite and back. This delay poses some problems for applications such as voice services and mobile telephony. Therefore, most mobile and voice services usually use LEO or MEO satellites to avoid the signal delays resulting from the inherent latency in GEO satellites. GEO satellites are usually used for broadcasting and data applications because of the larger area on the ground that they can cover.

Launching a satellite into space requires a very powerful multistage rocket to propel it into the right orbit. Satellite launch providers use proprietary rockets to launch satellites from sites such as the Kennedy Space Center at Cape Canaveral, Florida, the Baikonur Cosmodrome in Kazakhstan, Kourou in French Guiana, Vandenberg Air Force Base in California, Xichang in China, and Tanegashima Island in Japan. The U.S. space shuttle also has the ability to launch satellites.

Satellite communications use the very high-frequency range of 1–50 gigahertz (GHz; 1 gigahertz = 1,000,000,000 hertz) to transmit and receive signals. The frequency ranges or bands are identified by letters: (in order from low to high frequency) L-, S-, C-, X-, Ku-, Ka-, and V-bands. Signals in the lower range (L-, S-, and C-bands) of the satellite frequency spectrum are transmitted with low power, and thus larger antennas are needed to receive these signals. Signals in the higher end (X-, Ku-, Ka-, and V-bands) of this spectrum have more power; therefore, dishes as small as 45 cm (18 inches) in diameter can receive them. This makes the Ku-band and Ka-band spectrum ideal for direct-to-home (DTH) broadcasting, broadband data communications, and mobile telephony and data applications.

Satellite Applications

Advances in satellite technology have given rise to a healthy satellite services sector that provides various services to broadcasters, Internet service providers (ISPs), governments, the military and other sectors. There are three types of communication services that satellites provide: telecommunications, broadcasting, and data communications. Telecommunication services include telephone calls and services provided to telephone companies, as well as wireless, mobile, and cellular network providers.

Broadcasting services include radio and television delivered directly to the consumer and mobile broadcasting services. DTH, or satellite television, services (such as the DirecTV and DISH Network services in the United States) are received directly by households. Cable and network programming is delivered to local stations and affiliates largely via satellite. Satellites also play an important role in delivering programming to cell phones and other mobile devices, such as personal digital assistants and laptops.

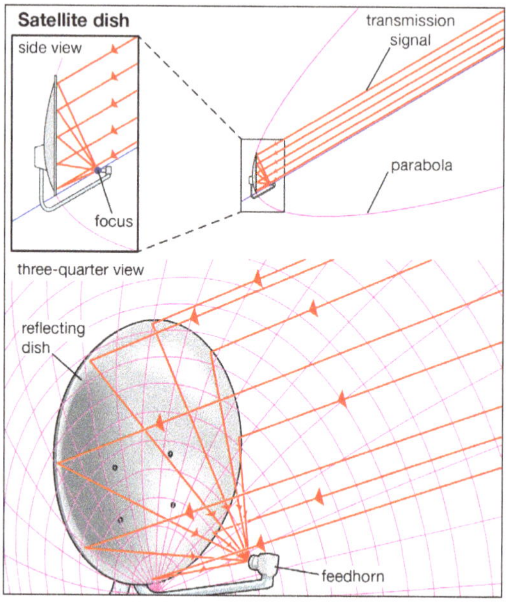

Parabolic satellite dish antenna.

Satellite dishes are often shaped like portions of a paraboloid (a parabola rotated about its central axis) in order to focus transmission signals onto the pickup receiver, or feedhorn. Typically, the section of the paraboloid used is offset from the centre so that the feedhorn and its support do not unduly block signals to the reflecting dish.

Data communications involve the transfer of data from one point to another. Corporations and organizations that require financial and other information to be exchanged between their various locations use satellites to facilitate the transfer of data through the use of very small-aperture terminal (VSAT) networks. With the growth of the Internet, a significant amount of Internet traffic goes through satellites, making ISPs one of the largest customers for satellite services.

Satellite communications technology is often used during natural disasters and emergencies when land-based communication services are down. Mobile satellite equipment can be deployed to disaster areas to provide emergency communication services.

One major technical disadvantage of satellites, particularly those in geostationary orbit, is an inherent delay in transmission. While there are ways to compensate for this delay, it makes some applications that require real-time transmission and feedback, such as voice communications, not ideal for satellites.

Satellites face competition from other media such as fibre optics, cable, and other land-based delivery systems such as microwaves and even power lines. The main advantage of satellites is that they can distribute signals from one point to many locations. As such, satellite technology is ideal for "point-to-multipoint" communications such as broadcasting. Satellite communication does not require massive investments on the ground—making it ideal for underserved and isolated areas with dispersed populations.

Satellites and other delivery mechanisms such as fibre optics, cable, and other terrestrial networks are not mutually exclusive. A combination of various delivery mechanisms may be needed, which has given rise to various hybrid solutions where satellites can be one of the links in the chain in combination with other media. Ground service providers called "teleports" have the capability to receive and transmit signals from satellites and also provide connectivity with other terrestrial networks.

Future of Satellite Communication

In a relatively short span of time, satellite technology has developed from the experimental to the sophisticated and powerful. Future communication satellites will have more on-board processing capabilities, more power, and larger-aperture antennas that will enable satellites to handle more bandwidth. Further improvements in satellites' propulsion and power systems will increase their service life to 20–30 years from the current 10–15 years. In addition, other technical innovations such as low-cost reusable launch vehicles are in development. With increasing video, voice, and data traffic requiring larger amounts of bandwidth, there is no dearth of emerging applications that will drive demand for the satellite services in the years to come. The demand for more bandwidth, coupled with the continuing innovation and development of satellite technology, will ensure the long-term viability of the commercial satellite industry well into the 21st century.

Near-field Communication

NFC stands for "Near Field Communication" and, as the name implies, it enables short-range communication between compatible devices. This requires at least one transmitting device, and another to receive the signal. A range of devices can use the NFC standard and will be considered either passive or active.

Passive NFC devices include tags, and other small transmitters, that can send information to other NFC devices without the need for a power source of their own. However, they don't process any information sent from other sources, and can't connect to other passive components. These often take the form of interactive signs on walls or advertisements.

Active devices are able to both send and receive data, and can communicate with each other as well as with passive devices. Smartphones are by far the most common form of active NFC device. Public transport card readers and touch payment terminals are also good examples of the technology.

Working of NFC

Just like Bluetooth and Wi-Fi, and all manner of other wireless signals, NFC works on the principle of sending information over radio waves. Near Field Communication is another standard for wireless data transitions. This means that devices must adhere to certain specifications in order to communicate with each other properly. The technology used in NFC is based on older RFID (Radio-frequency identification) ideas, which used electromagnetic induction in order to transmit information.

This marks the one major difference between NFC and Bluetooth/Wi-Fi. The former can be used to induce electric currents within passive components as well as just send data. This means that passive devices don't require their own power supply. They can instead be powered by the electromagnetic field produced by an active NFC component when it comes into range. Unfortunately, NFC technology does not command enough inductance to charge our smartphones, but QI wireless charging is based on the same principle.

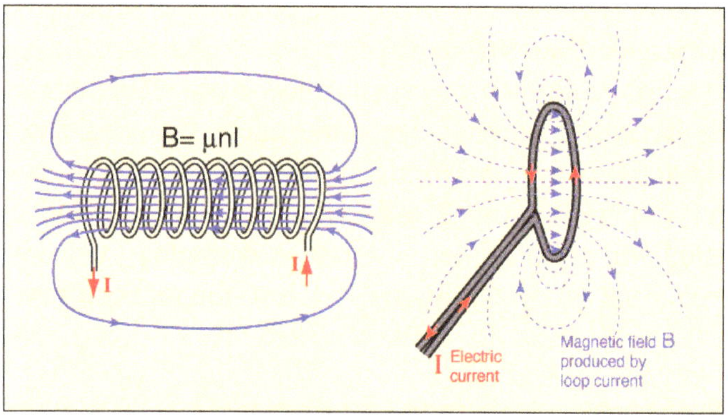

Electromagnetic fields can be used to transmit data or induce electrical currents in a receiving device. Passive NFC devices draw power from the fields produced by active devices, but the range is short.

The transmission frequency for data across NFC is 13.56 megahertz. You can send data at 106, 212, or 424 kilobits per second. That's quick enough for a range of data transfers — from contact details to swapping pictures and music.

To determine what sort of information will be exchanged between devices, the NFC standard currently has three distinct modes of operation. Perhaps the most common use in smartphones is the peer-to-peer mode. This allows two NFC-enabled devices to exchange various pieces of information between each other. In this mode, both devices switch between active when sending data and passive when receiving.

Read/write mode, on the other hand, is one-way data transmission. The active device, possibly your smartphone, links up with another device in order to read information from it. NFC advert tags use this mode.

The final mode of operation is card emulation. The NFC device can function as a smart or contactless credit card and make payments or tap into public transport systems.

Comparisons with Bluetooth

So how does NFC compare with other wireless technologies? You might think that NFC is a bit unnecessary, considering that Bluetooth has been more widely available for many years. However, there are several important technical differences between the two that gives NFC some significant benefits in certain circumstances. The major argument in favor of NFC is that it requires much less power consumption than Bluetooth. This makes NFC perfect for passive devices, such as the advertising tags as they can operate without a major power source.

However, this power-saving does have some major drawbacks. Most notably, the range of transmission is much shorter than Bluetooth. While NFC has a range of around 10 cm, just a few inches, Bluetooth connections can transmit data up to 10 meters or more from the source. Another drawback is that NFC is quite a bit slower than Bluetooth. It transmits data at a maximum speed of just 424 Kbit/s, compared to 2.1 Mbit/s with Bluetooth 2.1 or around 1 Mbit/s with Bluetooth Low Energy.

But NFC does have one major advantage: faster connectivity. Due to the use of inductive coupling, and the absence of manual pairing, it takes less than one-tenth of a second to establish a connection between two devices. While modern Bluetooth connects pretty fast, NFC is still super handy for certain scenarios. Namely mobile payments.

Samsung Pay, Android Pay, and even Apple Pay use NFC technology — though Samsung Pay works a bit differently than the others. While Bluetooth works better for connecting devices together for file transfers, sharing connections to speakers, and more, we anticipate that NFC will always have a place in this world thanks to mobile payments — a quickly expanding technology.

Mobile Telecommunications

Mobile phone or cellular telecommunications technology has been in widespread use since the early 1980s.

Since its first introduction, its use has increased very rapidly to the extent that a major portion of the global population has access to the technology. From developed nation to growing nation, mobile phone or cellular communications technology has been installed in all countries around the globe.

The cellular telecommunications industry has been a major driver in the growth of the radio and electronics industries.

Development of Cellular Communications

Although cellular communications are now accepted into everyday life, it took many years for their development to occur.

Although the basic concepts for cellular communications technology were proposed in the 1940s it was not until the mid-1980s that the radio technology and systems were deployed to enable widespread availability.

Usage of the cellular communications systems grew rapidly and as an example it was estimated that within the United Kingdom more calls were made using mobile phones than wired devices by 2011.

Another example of the growth of cellular telecommunications systems occurred in 2004 when the GSMA announced at Mobile World Congress in February 2004 that there were more than 1 billion GSM mobile subscribers – it had taken 12 years since the first network was launched. By comparison it had taken over 100 years for the same figure to be reached for wired telephone connections.

Then by 2015 more than 7 billion mobile subscriptions (for all technologies) were active. This is a major feat when it is realized that the global population was just over 7 billion. This meant that many people had more than one subscription, although market penetration was obviously very significant.

Cellular Telecommunications Generations

There is a lot of talk about the mobile phone generations. 3G moves on to 4G and then onto 5G.

Each mobile phone generation had its own aims and was able to provide different levels of functionality.

There may have also been several different competing standards within the different generations. For 3G cellular communications there were two main standards, but for 4G there was only one as there was global consensus on the system to use and this facilitated global roaming.

Generation	Approx launch year	Focus
1G	1979	Mobile voice
2G	1991	Mobile voice
3G	2001	Mobile Broadband
4G	2009	Mobile Broadband
5G	2020 (expected)	Ubiquitous connectivity

Key Cellular Communications Concepts

As the name indicates, cellular telecommunications technology is based around the concept of using a large number of base stations each covering a small area or cell. With each base station communicating with a reasonable number of users, it means that the whole system can accommodate a huge number of connections, and the levels of frequency use are good.

A cellular communications system has a number of different areas, each of which performs a different function. The main areas detailed below are the main ones that are normally referred to when discussing cellular communications systems. Each of these areas can often be split much further into different entities:

- Mobile handset or user equipment, UE: The user equipment or mobile phone is the element of a mobile communications system that the user sees. It connects to the network and enables the user to access voice and data services. The user equipment could also be a dongle used for accessing data on a laptop, or it could also be a modem on another form of device – for example cellular communications is starting to be used for Internet of Things, IoT applications and as a result it could be attached to a smart meter to automatically send meter readings or it could be used for any one of a host of other applications.
- Radio access network, RAN: The radio access network is at the periphery of the cellular communications system. It provides the link to the user equipment from the cellular network. It comprises a number of elements and broadly includes the base station and base station controller. With cellular communications technology advancing, the terms used and exactly what they contain is changing, but their basic function remains essentially the same.
- Core network: The core network is the hub of the cellular communications system. It manages the overall system as well as storing user data, manages access control, links to the external world and provides a host of other functions.

Wireless Location Tracking Technologies

GLONASS

GLONASS or Global Navigation Satellite System is the satellite navigation system developed by Russia that consists of 24 satellites, in three orbital planes, with eight satellites per plane. Russia started developing GLONASS in 1976 as an experimental military communications system. They launched the first GLONASS satellite in 1982 and the constellation became fully operational in 1995.

The satellites are placed into nominally circular orbits with target inclinations of 64.8 degrees and an orbital radius of 19,140 km; about 1,060 km lower than GPS satellites, with an orbit period of 11 hrs and 15 minutes.

Versions of GLONASS

- GLONASS: These satellites were launched in 1982 for the military and official organizations. They were intended to for weather, positioning, timing and velocity measurements.
- GLONASS-M: These satellites were launched in 2003 to add second civil code, which is important for GIS mapping receivers.
- GLONASS-K: These satellites were launched in 2011 to add third civil frequency. These are of 3 types - K1, K2 and KM.
- GLONASS-K2: These satellites will be launched after 2015 (currently in design phase).
- GLONASS-KM: These satellites will be launched after 2025 (currently in research phase).

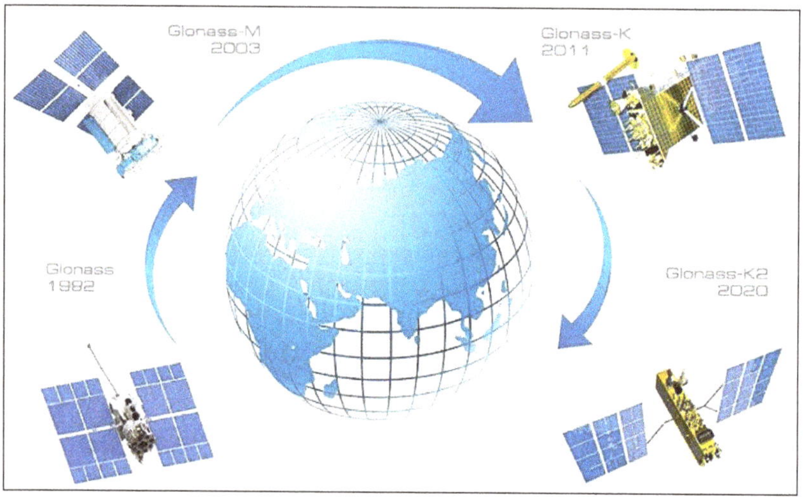

Currently, second generation GLONASS-M satellites as well as GLONASS-K1 satellites are in service while the GLONASS-K2 and KM satellites are under development. GLONASS signals have the same polarization (orientation of the electromagnetic waves) as GPS signals, and have comparable signal strength.

Each GLONASS satellite transmits a C/A-code for standard positioning on frequency L1, and a P-code for precise positioning on L1 and L2. The P-code is only available for military purposes. Unlike GPS and Galileo, GLONASS uses a different frequency for each satellite.

Indoor Positioning System

Like a GPS for indoor environments, IPS refers to the technology that helps locate people and objects indoors. That location information is then fed into some type of application software to make the information useful. For instance, IPS technologies enable a number of location-based solutions, including real-time location systems (RTLS), way-finding, inventory management, and first responder location systems.

There are a number of different technologies that can be used for indoor positioning, five of which are described here:

- Proximity-based Systems.
- WiFi-Based Systems.
- Ultra Wide-band Systems.
- Acoustic Systems.
- Infrared Systems.

Knowing how location tags work is like knowing what's happening "under the IPS hood." That's great, but it's how you use that location information that provides the real value—how will it make a difference in your business? The software it feeds into plays a big role in creating that value, so it's important to evaluate the system as a whole. If you're just starting out with indoor location tracking and have a tight budget for experimentation, buying a system that's expensive, hard to use, and difficult to scale is likely to negate the benefits you'll get from the collected data. On the other hand, a cost-effective, easy-to-use solution will help you make the most of your IPS investment, and create real business value as a result.

Types of Indoor Tracking Technology

Proximity-based Systems

Proximity-based systems can detect the general location of a person or object at room level within a facility. (That's in contrast to a precision system like ultra-wide-band, named below, which pinpoints the exact, precise location of something down to a "dot on the map.") Proximity-based systems use tags and beacons for indoor positioning, and they are either reader-based or reference point-based.

In a reader-based system, simple, inexpensive tags ("dumb" tags) transmit their identification continually to a number of reader devices. Those reader devices then pass the tags' identifications and signal strength to a backend system, which then calculates the positions of each tag.

A reference point-based system—of which AirFinder is a good example—uses standard Bluetooth Low Energy (BLE) beacons as location reference points together with "smart" location-aware tags. The tags calculate their own location based on the location of the reference points, and then connect to a central access point to relay this information. The BLE access points, spaced about every 100 feet in a facility, receive the encrypted location data from the tag and send it to the server.

Both types of proximity-based systems can accomplish simple indoor localization at the lowest cost. They are ubiquitous in healthcare and manufacturing, as well as some other industries. Reference point-based systems, however, are the most inexpensive of the proximity solutions because of their architecture—they don't need as many connected readers because of their low-cost reference points. They also enable prolonged battery life and deliver more accurate location information than reader-based systems.

WiFi-based Systems

In a WiFi-based system, tags are WiFi transmitters that send simple packets to a number of WiFi access points in a facility. These access points report the time and strength of that reading to a backend, which uses algorithms to compute position. The location information is then sent to the cloud.

WiFi indoor positioning systems give a fairly high level of accuracy—from three to five meters—because they use time difference of arrival (TDOA) measurements with wide bandwidth. But to achieve this level of accuracy you need at least three access points to "hear" each tag transmission.

Ultra Wide-band (UWB) Systems

UWB is a cool technology. Three or more ultra wide-band readers transmit a very wide pulse over a GHz of spectrum. The readers then listen for chirps from ultra wide-band tags. These tags have a spark-gap-style exciter that generates a little pulse within them, which creates a short, coded, very wide, nearly instantaneous burst. The readers then report very accurate time measurements from the tags back to a central server.

Because the UWB signal is extremely wide, the accuracy of the location information is very good—probably the most accurate of any system available. A drawback, however, is that UWB is the most expensive system to install. Even though UWB tags are inexpensive, every location has to have at least three (expensive) readers in it because of the limited range of the tags.

Acoustic Systems

A number of new indoor positioning systems have come onto the market that uses ultrasonic pulses from tags to locate them within an indoor environment. An acoustic system works almost exactly like UWB except it uses sound instead of radio. The tags emit a sound in the ultrasonic range (so you can't hear it). Receivers in the room (sometimes multiple, and sometimes a single "smart" one) pick up those sounds and locate the tags that way.

One benefit of using sound has to do with resolving multipath. If you're sending a transmission and taking a time measurement, you can guess the location based on the speed.

If that signal bounces off the wall on the way there, you now have a "multipath"—or maybe dozens of them. The ability to mathematically differentiate between a direct path and a multipath is purely a function of the speed of the medium divided by the bandwidth. So your multipath resolution ability is the speed of sound divided by the bandwidth. Acoustic systems require less signal bandwidth to resolve multipath because the speed of sound is so much less than the speed of light.

Sonar-based systems can be very accurate as well—even as accurate as UWB. The cost depends on your situation: If you're installing it in new construction, it won't cost much to wire a sensor into every room. If you're trying to retrofit an existing building with sensors, the cost would be high. Tags, however, are inexpensive.

For now, acoustic systems are an uncommonly-used, niche technology, though healthcare offers the most potential for use cases in the near future.

Infrared (IR) Systems

Infrared-based indoor localization systems use infrared light pulses (like a TV remote) to locate signals inside a building. IR receivers are installed in every room, and when the IR tag pulses, it is read by the IR receiver device.

Infrared is a near foolproof way to guarantee room-level accuracy. It uses light instead of radio waves, which can't go through walls—if the system says an asset is in room 4B, it is in room 4B without a doubt. Radio-based systems have more trouble with false positives, as the radio waves can sometimes be picked up by other readers through walls.

While the tags are low-cost and long-lasting, a drawback of infrared is that every room needs a wired IR reader to be installed in the ceiling. That's fine if you're installing it in new construction, but, just as with acoustic, retrofitting will be expensive. That's why infrared systems are commonly used in new hospital construction, where rooms are definitively segmented. In an open-space warehouse, infrared would be a challenge—if three receivers read a light pulse, there's no way to know which receiver the tag is closest to because it's difficult to measure the relative signal strength of infrared. Generally speaking, radio technologies work better in open spaces characteristic of warehouses and manufacturing facilities.

Satellite Navigation Systems

A GPS navigation system is a GPS receiver and audio/video (AV) components designed for a specific purpose such as a car-based or hand-held device or a smartphone app.

The global positioning system (GPS) is a 24-satellite navigation system that uses multiple satellite signals to find a receiver's position on earth. GPS was developed by the U.S. Department of Defense (DoD). The technology was originally used for military purposes. Since 1980, when GPS technology was made available to the consumer market, it has become common in cars, boats, cell phones, mobile devices and even personal heads-up display (HUD) glasses.

GPS receivers find their location by coordinating information from three or four satellite signals. That information includes the position of the satellite and the precise time of transmission. With three signals, any 2D position can be found on earth; additional satellite signals make it possible to find altitude.

GPS technology works in almost any condition and is accurate to within 3-15 meters, depending on the number of signals received, the spread of satellites in the sky and the technologies used in the receiver.

References

- Wireless-communication-introduction-types-applications: electronicshub.org, Retrieved 17 May, 2020
- Antenna-electronics, technology: britannica.com, Retrieved 09 July, 2020
- What-is-nfc-270730: androidauthority.com, Retrieved 28 April, 2020
- What-is-cellular-communications, connectivity-cellular-mobile-phone: electronics-notes.com, Retrieved 25 July, 2020
- Indoor-positioning-system: airfinder.com, Retrieved 10 August, 2020
- GPS-navigation-system: whatis.techtarget.com, Retrieved 05 June, 2020
- Satellite-communication, technology: britannica.com, Retrieved 28 May, 2020

2

Wireless Network and its Types

Wireless network uses radio waves to establish communication between computers and other network devices. A few of its types include wireless PAN, wireless ad hoc network, wireless LAN, wireless WAN, and global area network. This chapter sheds light on wireless network and its types for a thorough understanding of the subject.

Computer networks that are not connected by cables are called wireless networks. They generally use radio waves for communication between the network nodes. They allow devices to be connected to the network while roaming around within the network coverage.

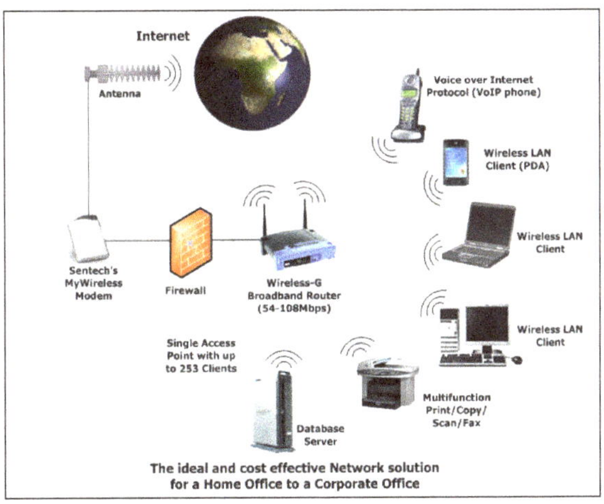

Types of Wireless Networks

- Wireless LANs: Connects two or more network devices using wireless distribution techniques.
- Wireless MANs: Connects two or more wireless LANs spreading over a metropolitan area.

- Wireless WANs: Connects large areas comprising LANs, MANs and personal networks.

Advantages of Wireless Networks

- It provides clutter-free desks due to the absence of wires and cables.
- It increases the mobility of network devices connected to the system since the devices need not be connected to each other.
- Accessing network devices from any location within the network coverage or Wi-Fi hotspot becomes convenient since laying out cables is not needed.
- Installation and setup of wireless networks are easier.
- New devices can be easily connected to the existing setup since they needn't be wired to the present equipment. Also, the number of equipment that can be added or removed to the system can vary considerably since they are not limited by the cable capacity. This makes wireless networks very scalable.
- Wireless networks require very limited or no wires. Thus, it reduces the equipment and setup costs.

Examples of Wireless Networks

- Mobile phone networks.
- Wireless sensor networks.
- Satellite communication networks.
- Terrestrial microwave networks.

Wireless Networking Standards

IEEE 802.11 is part of the IEEE 802 set of LAN protocols, and specifies the set of media access control (MAC) and physical layer (PHY) protocols for implementing wireless local area network (WLAN) Wi-Fi computer communication in various frequencies, including but not limited to 2.4 GHz, 5 GHz, and 60 GHz frequency bands.

They are the world's most widely used wireless computer networking standards, used in most home and office networks to allow laptops, printers, and smartphones to talk to each other and access the Internet without connecting wires. They are created and maintained by the Institute of Electrical and Electronics Engineers (IEEE) LAN/MAN Standards Committee (IEEE 802). The base version of the standard was released in 1997, and has had subsequent amendments. The standard and amendments provide the basis for wireless network products using the Wi-Fi brand. While each amendment is officially revoked when it is incorporated in the latest version of the standard, the corporate world tends to market to the revisions because they concisely denote capabilities of their products. As a result, in the marketplace, each revision tends to become its own standard.

The protocols are typically used in conjunction with IEEE 802.2, and are designed to interwork seamlessly with Ethernet, and are very often used to carry Internet Protocol traffic.

Although IEEE 802.11 specifications list channels that might be used, the radio frequency spectrum availability allowed varies significantly by regulatory domain.

This Wi-Fi router operates on the 2.4 GHz "G" standard, capable of transmitting 54 Mbit/s.

For comparison, this Netgear dual-band router from 2013 uses the "AC" standard, capable of transmitting 1900 Mbit/s.

The 802.11 family consists of a series of half-duplex over-the-air modulation techniques that use the same basic protocol. The 802.11 protocol family employ carrier-sense multiple access with collision avoidance whereby equipment listens to a channel for other users (including non 802.11 users) before transmitting each packet.

802.11-1997 was the first wireless networking standard in the family, but 802.11b was the first widely accepted one, followed by 802.11a, 802.11g, 802.11n, and 802.11ac.

Other standards in the family (c–f, h, j) are service amendments that are used to extend the current scope of the existing standard, which may also include corrections to a previous specification.

802.11b and 802.11g use the 2.4 GHz ISM band, operating in the United States under Part 15 of the U.S. Federal Communications Commission Rules and Regulations; 802.11n can also use that band. Because of this choice of frequency band, 802.11b/g/n equipment may occasionally suffer interference in the 2.4 GHz band from microwave ovens, cordless telephones, and Bluetooth devices etc. 802.11b and 802.11g control their interference and susceptibility to interference by using direct-sequence spread spectrum (DSSS) and orthogonal frequency-division multiplexing (OFDM) signaling methods, respectively.

802.11a uses the 5 GHz U-NII band, which, for much of the world, offers at least 23 non-overlapping 20 MHz-wide channels rather than the 2.4 GHz ISM frequency band offering only three non-overlapping 20 MHz-wide channels, where other adjacent channels overlap. Better or worse performance with higher or lower frequencies (channels) may be realized, depending on the environment. 802.11n can use either the 2.4 GHz or 5 GHz band; 802.11ac uses only the 5 GHz band.

The segment of the radio frequency spectrum used by 802.11 varies between countries. In the US, 802.11a and 802.11g devices may be operated without a license, as allowed in Part 15 of the FCC Rules and Regulations. Frequencies used by channels one through six of 802.11b and 802.11g fall within the 2.4 GHz amateur radio band. Licensed amateur radio operators may operate 802.11b/g devices under Part 97 of the FCC Rules and Regulations, allowing increased power output but not commercial content or encryption.

In 2018, the Wi-Fi Alliance began using a consumer-friendly generation numbering scheme for the publicly used 802.11 protocols. Wi-Fi generations 1–6 refer to the 802.11b, 802.11a, 802.11g, 802.11n, 802.11ac, and 802.11ax protocols, in that order.

802.11-1997 (802.11 Legacy)

The original version of the standard IEEE 802.11 was released in 1997 and clarified in 1999, but is now obsolete. It specified two net bit rates of 1 or 2 megabits per second (Mbit/s), plus forward error correction code. It specified three alternative physical layer technologies: diffuse infrared operating at 1 Mbit/s; frequency-hopping spread spectrum operating at 1 Mbit/s or 2 Mbit/s; and direct-sequence spread spectrum operating at 1 Mbit/s or 2 Mbit/s. The latter two radio technologies used microwave transmission over the Industrial Scientific Medical frequency band at 2.4 GHz. Some earlier WLAN technologies used lower frequencies, such as the U.S. 900 MHz ISM band.

Legacy 802.11 with direct-sequence spread spectrum was rapidly supplanted and popularized by 802.11b.

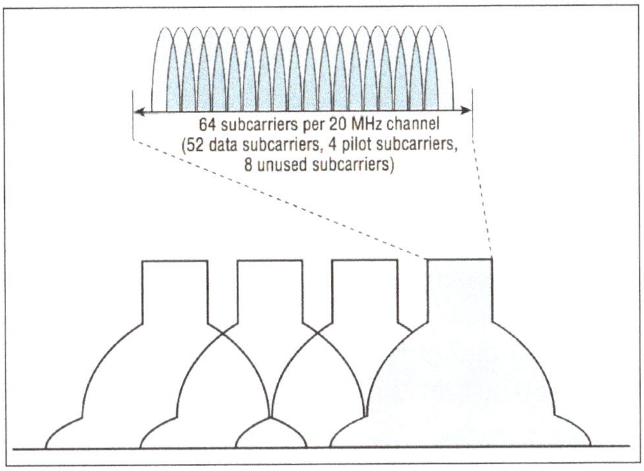

802.11a (OFDM Waveform)

802.11a, published in 1999, uses the same data link layer protocol and frame format as the original standard, but an OFDM based air interface (physical layer). It operates in the 5 GHz band with a maximum net data rate of 54 Mbit/s, plus error correction code, which yields realistic net achievable throughput in the mid-20 Mbit/s. It has seen widespread worldwide implementation, particularly within the corporate workspace.

Since the 2.4 GHz band is heavily used to the point of being crowded, using the relatively unused 5 GHz band gives 802.11a a significant advantage. However, this high carrier frequency also brings a disadvantage: the effective overall range of 802.11a is less than that of 802.11b/g. In theory, 802.11a signals are absorbed more readily by walls and other solid objects in their path due to their smaller wavelength, and, as a result, cannot penetrate as far as those of 802.11b. In practice, 802.11b typically has a higher range at low speeds (802.11b will reduce speed to 5.5 Mbit/s or even 1 Mbit/s at low signal strengths). 802.11a also suffers from interference, but locally there may be fewer signals to interfere with, resulting in less interference and better throughput.

802.11b

The 802.11b standard has a maximum raw data rate of 11 Mbit/s (Megabits per second), and uses the same media access method defined in the original standard. 802.11b products appeared on the market in early 2000, since 802.11b is a direct extension of the modulation technique defined in the original standard. The dramatic increase in throughput of 802.11b (compared to the original standard) along with simultaneous substantial price reductions led to the rapid acceptance of 802.11b as the definitive wireless LAN technology.

Devices using 802.11b experience interference from other products operating in the 2.4 GHz band. Devices operating in the 2.4 GHz range include microwave ovens, Bluetooth devices, baby monitors, cordless telephones, and some amateur radio equipment. As unlicensed intentional radiators in this ISM band, they must not interfere with and must tolerate interference from primary or secondary allocations (users) of this band, such as amateur radio.

802.11g

In June 2003, a third modulation standard was ratified: 802.11g. This works in the 2.4 GHz band (like 802.11b), but uses the same OFDM based transmission scheme as 802.11a. It operates at a maximum physical layer bit rate of 54 Mbit/s exclusive of forward error correction codes, or about 22 Mbit/s average throughputs. 802.11g hardware is fully backward compatible with 802.11b hardware, and therefore is encumbered with legacy issues that reduce throughput by ~21% when compared to 802.11a.

The then-proposed 802.11g standard was rapidly adopted in the market starting in January 2003, well before ratification, due to the desire for higher data rates as well as reductions in manufacturing costs. By summer 2003, most dual-band 802.11a/b products became dual-band/tri-mode, supporting a and b/g in a single mobile adapter card or access point. Details of making b and g work well together occupied much of the lingering technical process; in an 802.11g network, however, activity of an 802.11b participant will reduce the data rate of the overall 802.11g network.

Like 802.11b, 802.11g devices also suffer interference from other products operating in the 2.4 GHz band, for example wireless keyboards.

802.11-2007

In 2003, task group TGma was authorized to "roll up" many of the amendments to the 1999 version of the 802.11 standard. REVma or 802.11ma, as it was called, created a single document that merged 8 amendments (802.11a, b, d, e, g, h, i, j) with the base standard. Upon approval on March 8, 2007, 802.11REVma was renamed to the then-current base standard IEEE 802.11-2007.

802.11n

802.11n is an amendment that improves upon the previous 802.11 standards by adding multiple-input multiple-output antennas (MIMO). 802.11n operates on both the 2.4 GHz and the 5 GHz bands. Support for 5 GHz bands is optional. Its net data rate ranges from 54 Mbit/s to 600 Mbit/s. The IEEE has approved the amendment, and it was published in October 2009. Prior to the final ratification, enterprises were already migrating to 802.11n networks based on the Wi-Fi Alliance's certification of products conforming to a 2007 draft of the 802.11n proposal.

802.11-2012

In May 2007, task group TGmb was authorized to "roll up" many of the amendments to the 2007 version of the 802.11 standard. REVmb or 802.11mb, as it was called, created a single document that merged ten amendments (802.11k, r, y, n, w, p, z, v, u, s) with the 2007 base standard. In addition much cleanup was done, including a reordering of many of the clauses. Upon publication on March 29, 2012, the new standard was referred to as IEEE 802.11-2012.

802.11ac

IEEE 802.11ac-2013 is an amendment to IEEE 802.11, published in December 2013, that builds on 802.11n. Changes compared to 802.11n include wider channels (80 or 160 MHz versus 40 MHz) in the 5 GHz band, more spatial streams (up to eight versus four), higher-order modulation (up to 256-QAM vs. 64-QAM), and the addition of Multi-user MIMO (MU-MIMO). As of October 2013, high-end implementations support 80 MHz channels, three spatial streams, and 256-QAM, yielding a data rate of up to 433.3 Mbit/s per spatial stream, 1300 Mbit/s total, in 80 MHz channels in the 5 GHz band. Vendors have announced plans to release so-called "Wave 2" devices with support for 160 MHz channels, four spatial streams, and MU-MIMO in 2014 and 2015.

802.11ad

IEEE 802.11ad is an amendment that defines a new physical layer for 802.11 networks to operate in the 60 GHz millimeter wave spectrum. This frequency band has significantly different propagation characteristics than the 2.4 GHz and 5 GHz bands where Wi-Fi networks operate. Products implementing the 802.11ad standard are being brought to market under the WiGig brand name. The certification program is now being developed by the Wi-Fi Alliance instead of the now defunct Wireless Gigabit Alliance. The peak transmission rate of 802.11ad is 7 Gbit/s.

IEEE 802.11ad is a protocol used for very high data rates (about 8 Gbit/s) and for short range communication (about 1–10 meters).

TP-Link announced the world's first 802.11ad router in January 2016.

The WiGig standard is not too well known, although it was announced in 2009 and added to the IEEE 802.11 family in December 2012.

802.11af

IEEE 802.11af, also referred to as "White-Fi" and "Super Wi-Fi", is an amendment, approved in February 2014, that allows WLAN operation in TV white space spectrum in the VHF and UHF bands between 54 and 790 MHz. It uses cognitive radio technology to transmit on unused TV channels, with the standard taking measures to limit interference for primary users, such as analog TV, digital TV, and wireless microphones. Access points and stations determine their position using a satellite positioning system such as GPS, and use the Internet to query a geolocation database (GDB) provided by a regional regulatory agency to discover what frequency channels are available for use at a given time and position. The physical layer uses OFDM and is based on 802.11ac. The propagation path loss as well as the attenuation by materials such as brick and concrete is lower in the UHF and VHF bands than in the 2.4 GHz and 5 GHz bands, which increase the possible range. The frequency channels are 6 to 8 MHz wide, depending on the regulatory domain. Up to four channels may be bonded in either one or two contiguous blocks. MIMO operation is possible with up to four streams used for either space–time block code (STBC) or multi-user (MU) operation. The achievable data rate per spatial stream is 26.7 Mbit/s for 6 and 7 MHz channels, and 35.6 Mbit/s for 8 MHz channels. With four spatial streams and four bonded channels, the maximum data rate is 426.7 Mbit/s for 6 and 7 MHz channels and 568.9 Mbit/s for 8 MHz channels.

802.11-2016

IEEE 802.11-2016 which was known as IEEE 802.11 REVmc, is a revision based on IEEE 802.11-2012, incorporating 5 amendments (11ae, 11aa, 11ad, 11ac, 11af). In addition, existing MAC and PHY functions have been enhanced and obsolete features were removed or marked for removal. Some clauses and annexes have been renumbered.

802.11ah

IEEE 802.11ah, published in 2017, defines a WLAN system operating at sub-1 GHz license-exempt bands. Due to the favorable propagation characteristics of the low frequency spectra, 802.11ah can provide improved transmission range compared with the conventional 802.11 WLANs operating in the 2.4 GHz and 5 GHz bands. 802.11ah can be used for various purposes including large scale sensor networks, extended range hotspot, and outdoor Wi-Fi for cellular traffic offloading, whereas the available bandwidth is relatively narrow. The protocol intends consumption to be competitive with low power Bluetooth, at a much wider range.

802.11ai

IEEE 802.11ai is an amendment to the 802.11 standard that added new mechanisms for a faster initial link setup time.

802.11aj

IEEE 802.11aj is a rebanding of 802.11ad for use in the 45 GHz unlicensed spectrum available in some regions of the world (specifically China).

Alternatively known as China Milli-Meter Wave (CMMW).

802.11aq

IEEE 802.11aq is an amendment to the 802.11 standard that will enable pre-association discovery of services. This extends some of the mechanisms in 802.11u that enabled device discovery to further discover the services running on a device, or provided by a network.

802.11ax

IEEE 802.11ax (marketed as “Wi-Fi 6” by the Wi-Fi Alliance) is the successor to 802.11ac, and will increase the efficiency of WLAN networks. Currently in development, this project has the goal of providing 4x the throughput of 802.11ac at the user layer, having just 37% higher nominal data rates at the PHY layer. In the previous amendment of 802.11 (namely 802.11ac), Multi-User MIMO has been introduced, which is a spatial multiplexing technique. MU-MIMO allows the Access Point to form beams towards each Client, while transmitting information simultaneously. By doing so, the interference between Clients is reduced, and the overall throughput is increased, since multiple Clients can receive data at the same time. With 802.11ax, a similar multiplexing is introduced in the frequency domain, namely OFDMA. With this technique, multiple Clients are assigned with different Resource Units in the available spectrum. By doing so, an 80 MHz channel can be split into multiple Resource Units, so that multiple Clients receive different type of data over the same spectrum, simultaneously. In order to have enough amounts of subcarriers to support the requirements of OFDMA, the number of subcarriers is increased by a factor of 4 (compared to 802.11ac standard). In other words, for 20, 40, 80 and 160 MHz channels, there are 64, 128, 256 and 512 subcarriers in 802.11ac standard; while there are 256, 512, 1024 and 2048 subcarriers in 802.11ax standard. Since the available bandwidths have not changed and the number of subcarriers is increased with a factor of 4, the subcarrier spacing is reduced by a factor of 4 as well. This introduces 4 times longer OFDM symbols. E.g., for 802.11ac and 802.11ax, the duration of an OFDM symbol is 3.2 micro seconds and 12.8 micro seconds (both without guard intervals), respectively.

802.11ay

IEEE 802.11ay is a standard that is being developed. It is an amendment that defines a new physical layer for 802.11 networks to operate in the 60 GHz millimeter wave spectrum. It will be an extension of the existing 11ad, aimed to extend the throughput, range and use-cases. The main use-cases include: indoor operation, out-door back-haul and short range communications. The peak transmission rate of 802.11ay is 20 Gbit/s. The main extensions include: channel bonding (2, 3 and 4), MIMO and higher modulation schemes.

802.11be

IEEE 802.11be Extremely High Throughput (EHT) is the potential next amendment of the 802.11 IEEE standards. It will build upon 802.11ax, focusing on WLAN indoor and outdoor operation with stationary and pedestrian speeds in the 2.4 GHz, 5 GHz, and 6 GHz frequency bands. Being the potential successor of Wi-Fi 6, the Wi-Fi Alliance will most likely certify it as Wi-Fi 7.

Common Misunderstandings about Achievable Throughput

Across all variations of 802.11, maximum achievable throughputs are given either based on measurements under ideal conditions or in the layer-2 data rates. However, this does not apply to typical deployments in which data is being transferred between two endpoints, of which at least one is typically connected to a wired infrastructure and the other endpoint is connected to an infrastructure via a wireless link.

This means that, typically, data frames pass an 802.11 (WLAN) medium, and are being converted to 802.3 (Ethernet) or vice versa. Due to the difference in the frame (header) lengths of these two media, the application's packet size determines the speed of the data transfer. This means applications that use small packets (e.g., VoIP) create data-flows with high-overhead traffic (i.e., a low goodput). Other factors that contribute to the overall application data rate are the speed with which the application transmits the packets (i.e., the data rate) and, of course, the energy with which the wireless signals is received. The latter is determined by distance and by the configured output power of the communicating devices.

The same references apply to the attached graphs that show measurements of UDP throughput. Each represents an average (UDP) throughput (please note that the error bars are there, but barely visible due to the small variation) of 25 measurements. Each is with a specific packet size (small or large) and with a specific data rate (10 Kbit/s – 100 Mbit/s). Markers for traffic profiles of common applications are included as well. These figures assume there are no packet errors, which if occurring will lower transmission rate further.

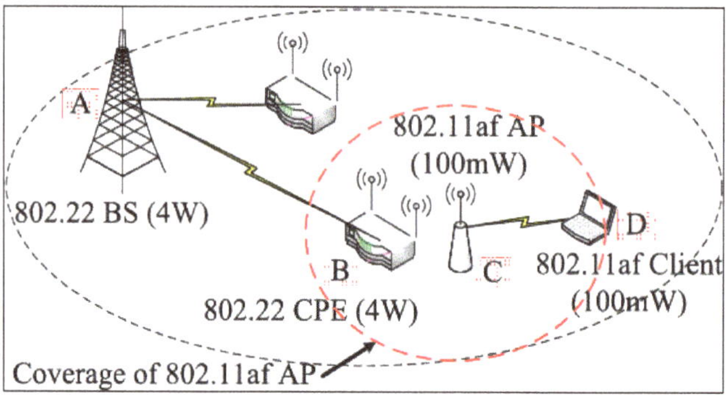

Channels and Frequencies

802.11b, 802.11g, and 802.11n-2.4 utilize the 2.400–2.500 GHz spectrum, one of the ISM bands. 802.11a, 802.11n and 802.11ac use the more heavily regulated 4.915–5.825 GHz band. These are commonly referred to as the "2.4 GHz and 5 GHz bands" in most sales literature. Each spectrum is sub-divided into channels with a center frequency and bandwidth, analogous to the way radio and TV broadcast bands are sub-divided.

The 2.4 GHz band is divided into 14 channels spaced 5 MHz apart, beginning with channel 1, which is centered on 2.412 GHz. The latter channels have additional restrictions or are unavailable for use in some regulatory domains.

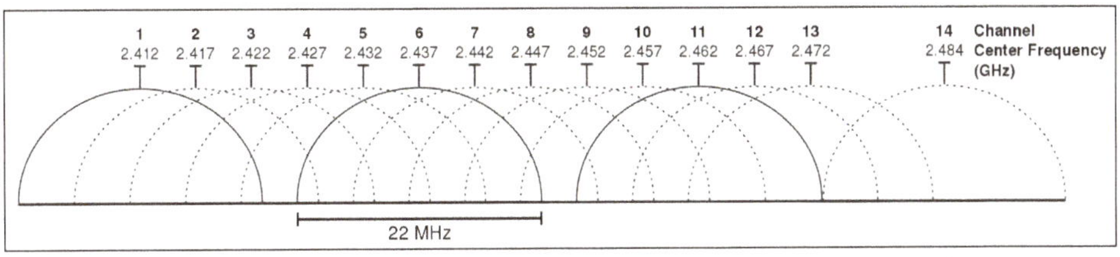

Graphical representation of Wi-Fi channels in the 2.4 GHz band.

The channel numbering of the 5.725–5.875 GHz spectrum is less intuitive due to the differences in regulations between countries.

Channel Spacing within the 2.4 Ghz Band

In addition to specifying the channel center frequency, 802.11 also specify (in Clause 17) a spectral mask defining the permitted power distribution across each channel. The mask requires the signal be attenuated a minimum of 20 dB from its peak amplitude at ±11 MHz from the center frequency, the point at which a channel is effectively 22 MHz wide. One consequence is that stations can use only every fourth or fifth channel without overlap.

Availability of channels is regulated by country, constrained in part by how each country allocates radio spectrum to various services. At one extreme, Japan permits the use of all 14 channels for 802.11b, and 1–13 for 802.11g/n-2.4. Other countries such as Spain initially allowed only channels 10 and 11, and France allowed only 10, 11, 12, and 13; however, Europe now allows channels 1 through 13. North America and some Central and South American countries allow only 1 through 11.

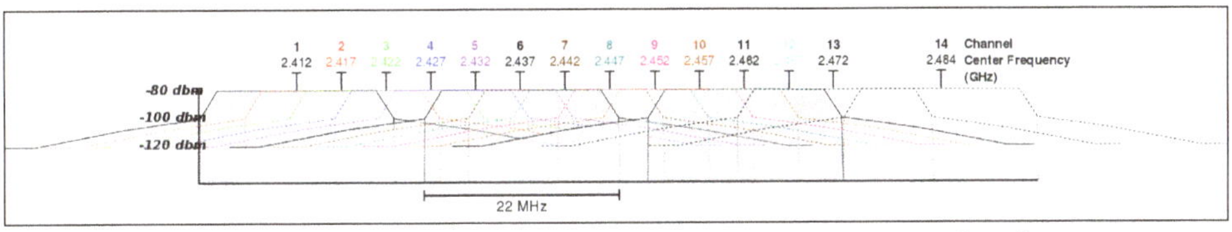

Spectral masks for 802.11g channels 1–14 in the 2.4 GHz band.

Since the spectral mask defines only power output restrictions up to ±11 MHz from the center frequency to be attenuated by −50 dBr, it is often assumed that the energy of the channel extends no further than these limits. It is more correct to say that, given the separation between channels, the overlapping signal on any channel should be sufficiently attenuated to minimally interfere with a transmitter on any other channel. Due to the near-far problem a transmitter can impact (desense) a receiver on a "non-overlapping" channel, but only if it is close to the victim receiver (within a meter) or operating above allowed power levels. Conversely, a sufficiently distant transmitter on an overlapping channel can have little to no significant effect.

Confusion often arises over the amount of channel separation required between transmitting devices. 802.11b was based on direct-sequence spread spectrum (DSSS) modulation and utilized a channel bandwidth of 22 MHz, resulting in three "non-overlapping" channels (1, 6, and 11). 802.11g was based on OFDM modulation and utilized a channel bandwidth of 20 MHz. This occasionally leads to the belief that four "non-overlapping" channels (1, 5, 9, and 13) exist under 802.11g, although this is not the case as per 17.4.6.3 Channel Numbering of operating channels of the IEEE Std 802.11, which states "In a multiple cell network topology, overlapping and/or adjacent cells using different channels can operate simultaneously without interference if the distance between the center frequencies is at least 25 MHz".

This does not mean that the technical overlap of the channels recommends the non-use of overlapping channels. The amount of inter-channel interference seen on a configuration using channels 1, 5, 9, and 13 (which is permitted in Europe, but not in North America) is barely different from a three-channel configuration, but with an entire extra channel.

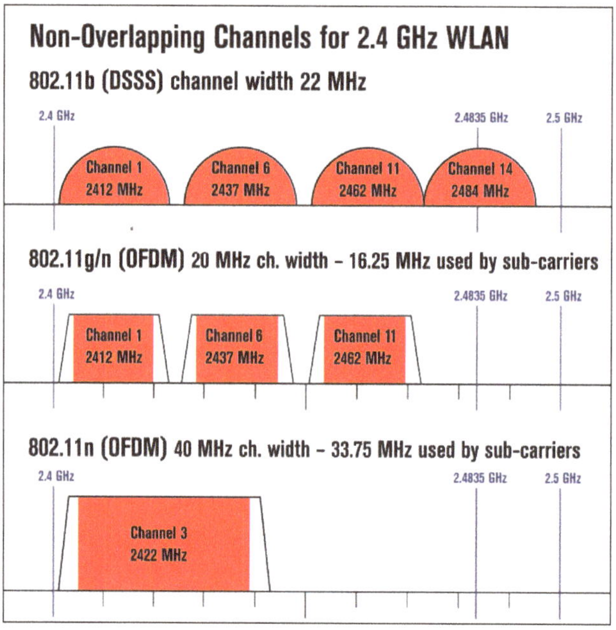

802.11 non-overlapping channels for 2.4GHz. Covers 802.11b,g,n.

However, overlap between channels with more narrow spacing (e.g. 1, 4, 7, 11 in North America) may cause unacceptable degradation of signal quality and throughput, particularly when users transmit near the boundaries of AP cells.

Regulatory Domains and Legal Compliance

IEEE uses the phrase regdomain to refer to a legal regulatory region. Different countries define different levels of allowable transmitter power, time that a channel can be occupied, and different available channels. Domain codes are specified for the United States, Canada, ETSI (Europe), Spain, France, Japan, and China.

Most Wi-Fi certified devices default to regdomain 0, which means least common denominator settings, i.e., the device will not transmit at a power above the allowable power in any nation, nor will it use frequencies that are not permitted in any nation.

The regdomain setting is often made difficult or impossible to change so that the end users do not conflict with local regulatory agencies such as the United States' Federal Communications Commission.

Layer 2 – Datagrams

The datagrams are called frames. Current 802.11 standards specify frame types for use in transmission of data as well as management and control of wireless links.

Frames are divided into very specific and standardized sections. Each frame consists of a MAC header, payload, and frame check sequence (FCS). Some frames may not have a payload.

Field	Frame control	Duration id	Address 1	Address 2	Address 3	Sequence control	Address 4	QoS control	HT control	Frame body	Frame check sequence
Length	2	2	6	6	6	0, or 2	6	0, or 2	0, or 4	*Variable*	4

The first two bytes of the MAC header form a frame control field specifying the form and function of the frame. This frame control field is subdivided into the following subfields:

- Protocol Version: Two bits representing the protocol version. Currently used protocol version is zero. Other values are reserved for future use.
- Type: Two bits identifying the type of WLAN frame. Control, Data, and Management are various frame types defined in IEEE 802.11.
- Subtype: Four bits providing additional discrimination between frames. Type and Subtype are used together to identify the exact frame.
- ToDS and FromDS: Each is one bit in size. They indicate whether a data frame is headed for a distribution system. Control and management frames set these values to zero. All the data frames will have one of these bits set. However communication within an independent basic service set (IBSS) network always set these bits to zero.
- More Fragments: The More Fragments bit is set when a packet is divided into multiple frames for transmission. Every frame except the last frame of a packet will have this bit set.
- Retry: Sometimes frames require retransmission, and for this there is a Retry bit that is set to one when a frame is resent. This aids in the elimination of duplicate frames.

- Power Management: This bit indicates the power management state of the sender after the completion of a frame exchange. Access points are required to manage the connection, and will never set the power-saver bit.
- More Data: The More Data bit is used to buffer frames received in a distributed system. The access point uses this bit to facilitate stations in power-saver mode. It indicates that at least one frame is available, and addresses all stations connected.
- Protected Frame: The Protected Frame bit is set to one if the frame body is encrypted by a protection mechanism such as Wired Equivalent Privacy (WEP), Wi-Fi Protected Access (WPA), or Wi-Fi Protected Access II (WPA2).
- Order: This bit is set only when the "strict ordering" delivery method is employed. Frames and fragments are not always sent in order as it causes a transmission performance penalty.

The next two bytes are reserved for the Duration ID field. This field can take one of three forms: Duration, Contention-Free Period (CFP), and Association ID (AID).

An 802.11 frame can have up to four address fields. Each field can carry a MAC address. Address 1 is the receiver, Address 2 is the transmitter, Address 3 is used for filtering purposes by the receiver. Address 4 is only present in data frames transmitted between access points in an Extended Service Set or between intermediate nodes in a mesh network.

The remaining fields of the header are:

- The Sequence Control field is a two-byte section used for identifying message order as well as eliminating duplicate frames. The first 4 bits are used for the fragmentation number, and the last 12 bits are the sequence number.
- An optional two-byte Quality of Service control field, present in QoS Data frames; it was added with 802.11e.

The payload or frame body field is variable in size, from 0 to 2304 bytes plus any overhead from security encapsulation, and contains information from higher layers.

The Frame Check Sequence (FCS) is the last four bytes in the standard 802.11 frame. Often referred to as the Cyclic Redundancy Check (CRC), it allows for integrity check of retrieved frames. As frames are about to be sent, the FCS is calculated and appended. When a station receives a frame, it can calculate the FCS of the frame and compare it to the one received. If they match, it is assumed that the frame was not distorted during transmission.

Management Frames

Management frames are not always authenticated, and allow for the maintenance, or discontinuance, of communication. Some common 802.11 subtypes include:

- Authentication frame: 802.11 authentication begins with the wireless network interface card (WNIC) sending an authentication frame to the access point containing its identity. With an open system authentication, the WNIC sends only a single authentication frame, and the access point responds with an authentication frame of its own indicating acceptance or rejection. With shared key authentication, after the WNIC sends its initial authentication request it will receive an authentication frame from the access point containing challenge text. The WNIC sends an authentication frame containing the encrypted version of the challenge text to the access point. The access point ensures the text was encrypted with the correct key by decrypting it with its own key. The result of this process determines the WNIC's authentication status.

- Association request frame: Sent from a station it enables the access point to allocate resources and synchronize. The frame carries information about the WNIC, including supported data rates and the SSID of the network the station wishes to associate with. If the request is accepted, the access point reserves memory and establishes an association ID for the WNIC.
- Association response frame: Sent from an access point to a station containing the acceptance or rejection to an association request. If it is an acceptance, the frame will contain information such an association ID and supported data rates.
- Beacon frame: Sent periodically from an access point to announce its presence and provide the SSID, and other parameters for WNICs within range.
- Deauthentication frame: Sent from a station wishing to terminate connection from another station.
- Disassociation frame: Sent from a station wishing to terminate connection. It's an elegant way to allow the access point to relinquish memory allocation and remove the WNIC from the association table.
- Probe request frame: Sent from a station when it requires information from another station.
- Probe response frame: Sent from an access point containing capability information, supported data rates, etc., after receiving a probe request frame.
- Reassociation request frame: A WNIC sends a reassociation request when it drops from range of the currently associated access point and finds another access point with a stronger signal. The new access point coordinates the forwarding of any information that may still be contained in the buffer of the previous access point.
- Reassociation response frame: Sent from an access point containing the acceptance or rejection to a WNIC reassociation request frame. The frame includes information required for association such as the association ID and supported data rates.

- Action frame: extending management frame to control certain action. Some of action categories are Block Ack, Radio Measurement, Fast BSS Transition, etc. These frames are sent by a station when it needs to tell its peer for certain action to be taken. For example, a station can tell another station to set up a block acknowledgement by sending an ADDBA Request action frame. The other station would then respond with an ADDBA Response action frame.

The body of a management frame consists of frame-subtype-dependent fixed fields followed by a sequence of information elements (IEs).

The common structure of an IE is as follows:

Field	Type	Length	Data
Length	1	1	1–252

Control Frames

Control frames facilitate in the exchange of data frames between stations. Some common 802.11 control frames include:

- Acknowledgement (ACK) frame: After receiving a data frame, the receiving station will send an ACK frame to the sending station if no errors are found. If the sending station doesn't receive an ACK frame within a predetermined period of time, the sending station will resend the frame.
- Request to Send (RTS) frame: The RTS and CTS frames provide an optional collision reduction scheme for access points with hidden stations. A station sends a RTS frame as the first step in a two-way handshake required before sending data frames.
- Clear to Send (CTS) frame: A station responds to an RTS frame with a CTS frame. It provides clearance for the requesting station to send a data frame. The CTS provides collision control management by including a time value for which all other stations are to hold off transmission while the requesting station transmits.

Data Frames

Data frames carry packets from web pages, files, etc. within the body. The body begins with an IEEE 802.2 header, with the Destination Service Access Point (DSAP) specifying the protocol, followed by a Subnetwork Access Protocol (SNAP) header if the DSAP is hex AA, with the organizationally unique identifier (OUI) and protocol ID (PID) fields specifying the protocol. If the OUI is all zeroes, the protocol ID field is an EtherType value. Almost all 802.11 data frames use 802.2 and SNAP headers, and most use an OUI of 00:00:00 and an EtherType value.

Similar to TCP congestion control on the internet, frame loss is built into the operation of 802.11. To select the correct transmission speed or Modulation and Coding Scheme, a rate control algorithm may test different speeds. The actual packet loss rate of an Access points vary widely for different link conditions. There are variations in the loss rate experienced on production Access points, between 10% and 80%, with 30% being a common average. It is important to be aware that the link layer should recover these lost frames. If the sender does not receive an Acknowledgement (ACK) frame, then it will be resent.

Security

In 2001, a group from the University of California, Berkeley presented a paper describing weaknesses in the 802.11 Wired Equivalent Privacy (WEP) security mechanism defined in the original standard; they were followed by Fluhrer, Mantin, and Shamir's paper titled "Weaknesses in the Key Scheduling Algorithm of RC4". Not long after, Adam Stubblefield and AT&T publicly announced the first verification of the attack. In the attack, they were able to intercept transmissions and gain unauthorized access to wireless networks.

The IEEE set up a dedicated task group to create a replacement security solution, 802.11i (previously this work was handled as part of a broader 802.11e effort to enhance the MAC layer). The Wi-Fi Alliance announced an interim specification called Wi-Fi Protected Access (WPA) based on a subset of the then current IEEE 802.11i draft. These started to appear in products in mid-2003. IEEE 802.11i (also known as WPA2) itself was ratified in June 2004, and uses the Advanced Encryption Standard AES, instead of RC4, which was used in WEP. The modern recommended encryption for the home/consumer space is WPA2 (AES Pre-Shared Key), and for the enterprise space is WPA2 along with a RADIUS authentication server (or another type of authentication server) and a strong authentication method such as EAP-TLS.

In January 2005, the IEEE set up yet another task group "w" to protect management and broadcast frames, which previously were sent unsecured. Its standard was published in 2009.

In December 2011, a security flaw was revealed that affects some wireless routers with a specific implementation of the optional Wi-Fi Protected Setup (WPS) feature. While WPS is not a part of 802.11, the flaw allows an attacker within the range of the wireless router to recover the WPS PIN and, with it, the router's 802.11i password in a few hours.

In late 2014, Apple announced that its iOS 8 mobile operating system would scramble MAC addresses during the pre-association stage to thwart retail footfall tracking made possible by the regular transmission of uniquely identifiable probe requests.

Wi-Fi users may be subjected to a Wi-Fi deauthentication attack to eavesdrop, attack passwords or simply to force the use of another, usually more expensive access point.

Non-standard 802.11 Extensions and Equipment

Many companies implement wireless networking equipment with non-IEEE standard 802.11 extensions either by implementing proprietary or draft features. These changes may lead to incompatibilities between these extensions.

Wireless Application Protocol

Wireless Application Protocol (WAP) is a technical standard for accessing information over a mobile wireless network. A WAP browser is a web browser for mobile devices such as mobile phones that uses the protocol. Introduced in 1999, WAP achieved some popularity in the early 2000s, but by the 2010s it had been largely superseded by more modern standards. Most modern handset internet browsers now fully support HTML, so they do not need to use WAP markup for web page compatibility, and therefore, most are no longer able to render and display pages written in WML, WAP's markup language.

Before the introduction of WAP, mobile service providers had limited opportunities to offer interactive data services, but needed interactivity to support Internet and Web applications such as email, stock prices, news and sports headlines. The Japanese i-mode system offered another major competing wireless data protocol.

Technical Specifications

WAP Protocol Stack

The WAP standard described a protocol suite or stack allowing the interoperability of WAP equipment and software with different network technologies, such as GSM and IS-95 (also known as CDMA).

Wireless Application Environment (WAE)	WAP protocol suite
Wireless Session Protocol (WSP)	
Wireless Transaction Protocol (WTP)	
Wireless Transport Layer Security (WTLS)	
Wireless Datagram Protocol (WDP)	
*** Any wireless data network ***	

The bottom-most protocol in the suite, the Wireless Datagram Protocol (WDP), functions as an adaptation layer that makes every data network look a bit like UDP to the upper layers by providing unreliable transport of data with two 16-bit port numbers (origin and destination). All the upper layers view WDP as one and the same protocol, which has several "technical realizations" on top of other "data bearers" such as SMS, USSD, etc. On native IP bearers such as GPRS, UMTS packet-radio service, or PPP on top of a circuit-switched data connection, WDP is in fact exactly UDP.

WTLS, an optional layer, provides a public-key cryptography-based security mechanism similar to TLS.

WTP provides transaction support (reliable request/response) adapted to the wireless world. WTP supports more effectively than TCP the problem of packet loss, which occurs commonly in 2G wireless technologies in most radio conditions, but is misinterpreted by TCP as network congestion.

This protocol suite allows a terminal to transmit requests that have an HTTP or HTTPS equivalent to a WAP gateway; the gateway translates requests into plain HTTP.

The Wireless Application Environment (WAE) space defines application-specific markup languages.

For WAP version 1.X, the primary language of the WAE is Wireless Markup Language (WML). In WAP 2.0, the primary markup language is XHTML Mobile Profile.

WAP Push

WAP Push was incorporated into the specification to allow the WAP content to be pushed to the mobile handset with minimal user intervention. A WAP Push is basically a specially encoded message which includes a link to a WAP address.

WAP Push was specified on top of Wireless Datagram Protocol (WDP); as such, it can be delivered over any WDP-supported bearer, such as GPRS or SMS. Most GSM networks have a wide range of modified processors, but GPRS activation from the network is not generally supported, so WAP Push messages have to be delivered on top of the SMS bearer.

On receiving a WAP Push, a WAP 1.2 (or later) -enabled handset will automatically give the user the option to access the WAP content. This is also known as WAP Push SI (Service Indication). A variant, known as WAP Push SL (Service Loading), directly opens the browser to display the WAP content, without user interaction. Since this behaviour raises security concerns, some handsets handle WAP Push SL messages in the same way as SI, by providing user interaction.

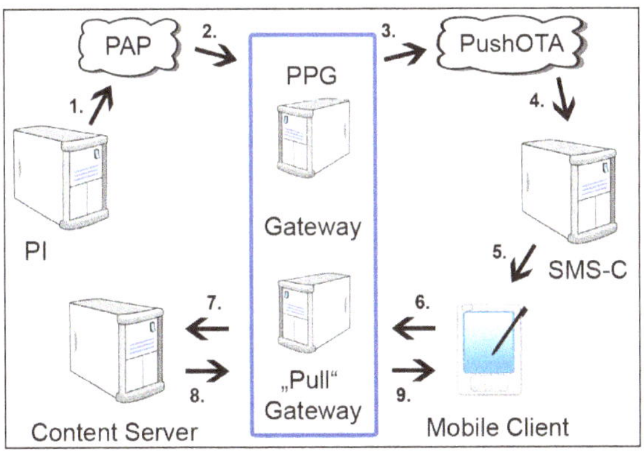

WAP push process.

The network entity that processes WAP Pushes and delivers them over an IP or SMS Bearer is known as a Push Proxy Gateway (PPG).

WAP 2.0

A re-engineered 2.0 version was released in 2002. It uses a cut-down version of XHTML with end-to-end HTTP, dropping the gateway and custom protocol suite used to communicate with it. A WAP gateway can be used in conjunction with WAP 2.0; however, in this scenario, it is used as a standard proxy server. The WAP gateway's role would then shift from one of translation to adding additional information to each request. This would be configured by the operator and could include telephone numbers, location, billing information, and handset information.

Mobile devices process XHTML Mobile Profile (XHTML MP), the markup language defined in WAP 2.0. It is a subset of XHTML and a superset of XHTML Basic. A version of Cascading Style Sheets (CSS) called WAP CSS is supported by XHTML MP.

MMS

Multimedia Messaging Service (MMS) is a combination of WAP and SMS allowing for sending of picture messages.

Wireless Sensor Network

Wireless sensor network (WSN) refers to a group of spatially dispersed and dedicated sensors for monitoring and recording the physical conditions of the environment and organizing the collected data at a central location. WSNs measure environmental conditions like temperature, sound, pollution levels, humidity, wind, and so on.

These are similar to wireless ad hoc networks in the sense that they rely on wireless connectivity and spontaneous formation of networks so that sensor data can be transported wirelessly. WSNs are spatially distributed autonomous sensors to monitor physical or environmental conditions, such as temperature, sound, pressure, etc. and to cooperatively pass their data through the network to a main location. The more modern networks are bi-directional, also enabling control of sensor activity. The development of wireless sensor networks was motivated by military applications such as battlefield surveillance; today such networks are used in many industrial and consumer applications, such as industrial process monitoring and control, machine health monitoring, and so on.

The WSN is built of "nodes" – from a few to several hundreds or even thousands, where each node is connected to one (or sometimes several) sensors. Each such sensor network node has typically several parts: a radio transceiver with an internal antenna or connection to an external antenna, a microcontroller, an electronic circuit for interfacing with the sensors and an energy source, usually a battery or an embedded form of energy harvesting. A sensor node might vary in size from that of a shoebox down to the size of a grain of dust, although functioning "motes" of genuine microscopic dimensions have yet to be created. The cost of sensor nodes is similarly variable, ranging from a few to hundreds of dollars, depending on the complexity of the individual sensor nodes. Size and cost constraints on sensor nodes result in corresponding constraints on resources such as energy, memory, computational speed and communications bandwidth. The topology of the WSNs can vary from a simple star network to an advanced multi-hop wireless mesh network. The propagation technique between the hops of the network can be routing or flooding.

In computer science and telecommunications, wireless sensor networks are an active research area with numerous workshops and conferences arranged each year, for example IPSN, SenSys, and EWSN. As of 2010, wireless sensor networks have reached approximately 120 million remote units worldwide.

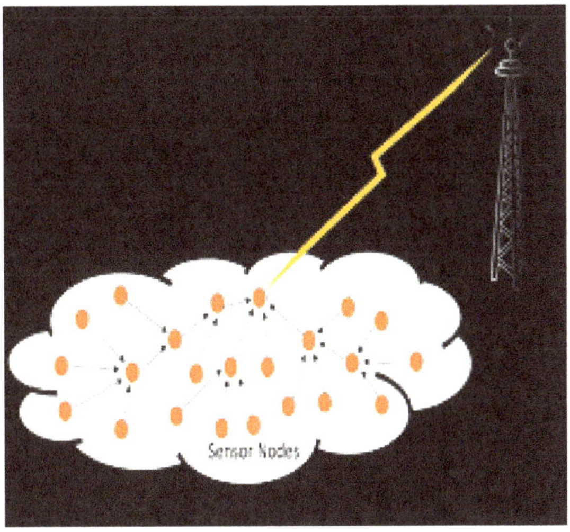

Representation of a wireless sensor network.

Applications

Area Monitoring

Area monitoring is a common application of WSNs. In area monitoring, the WSN is deployed over a region where some phenomenon is to be monitored. A military example is the use of sensors to detect enemy intrusion; a civilian example is the geo-fencing of gas or oil pipelines.

Health Care Monitoring

There are several types of sensor networks for medical applications: Implanted, wearable, and environment-embedded. Implantable medical devices are those that are inserted inside the human body. Wearable devices are used on the body surface of a human or just at close proximity of the user. Environment-embedded systems employ sensors contained in the environment. Possible applications include body position measurement, location of persons, overall monitoring of ill patients in hospitals and at home. Devices embedded in the environment track the physical state of a person for continuous health diagnosis, using as input the data from a network of depth cameras, a sensing floor, or other similar devices. Body-area networks can collect information about an individual's health, fitness, and energy expenditure. In health care applications the privacy and authenticity of user data has prime importance. Especially due to the integration of sensor networks, with IoT, the user authentication becomes more challenging; however, a solution is presented in recent work.

Environmental/Earth Sensing

There are many applications in monitoring environmental parameters, examples of which are given below. They share the extra challenges of harsh environments and reduced power supply.

Air Pollution Monitoring

Wireless sensor networks have been deployed in several cities (Stockholm, London, and Brisbane) to monitor the concentration of dangerous gases for citizens. These can take advantage of the ad hoc wireless links rather than wired installations, which also make them more mobile for testing readings in different areas.

Forest Fire Detection

A network of Sensor Nodes can be installed in a forest to detect when a fire has started. The nodes can be equipped with sensors to measure temperature, humidity and gases which are produced by fire in the trees or vegetation. The early detection is crucial for a successful action of the firefighters; thanks to Wireless Sensor Networks, the fire brigade will be able to know when a fire is started and how it is spreading.

Landslide Detection

A landslide detection system makes use of a wireless sensor network to detect the slight movements of soil and changes in various parameters that may occur before or during a landslide. Through the data gathered it may be possible to know the impending occurrence of landslides long before it actually happens.

Water Quality Monitoring

Water quality monitoring involves analyzing water properties in dams, rivers, lakes and oceans, as well as underground water reserves. The use of many wireless distributed sensors enables the creation of a more accurate map of the water status, and allows the permanent deployment of monitoring stations in locations of difficult access, without the need of manual data retrieval.

Natural Disaster Prevention

Wireless sensor networks can be effective in preventing adverse consequences of natural disasters, like floods. Wireless nodes have been deployed successfully in rivers, where changes in water levels must be monitored in real time.

Industrial Monitoring

Machine Health Monitoring

Wireless sensor networks have been developed for machinery condition-based maintenance (CBM) as they offer significant cost savings and enable new functionality.

Wireless sensors can be placed in locations difficult or impossible to reach with a wired system, such as rotating machinery and untethered vehicles.

Data Logging

Wireless sensor networks also are used for the collection of data for monitoring of environmental information. This can be as simple as monitoring the temperature in a fridge or the level of water in overflow tanks in nuclear power plants. The statistical information can then be used to show how systems have been working. The advantage of WSNs over conventional loggers is the "live" data feed that is possible.

Water/Waste Water Monitoring

Monitoring the quality and level of water includes many activities such as checking the quality of underground or surface water and ensuring a country's water infrastructure for the benefit of both human and animal. It may be used to protect the wastage of water.

Structural Health Monitoring

Wireless sensor networks can be used to monitor the condition of civil infrastructure and related geo-physical processes close to real time, and over long periods through data logging, using appropriately interfaced sensors.

Wine Production

Wireless sensor networks are used to monitor wine production, both in the field and the cellar.

Threat Detection

The Wide Area Tracking System (WATS) is a prototype network for detecting a ground-based nuclear device such as a nuclear "briefcase bomb." WATS is being developed at the Lawrence Livermore National Laboratory (LLNL). WATS would be made up of wireless gamma and neutron sensors connected through a communications network. Data picked up by the sensors undergoes "data fusion", which converts the information into easily interpreted forms; this data fusion is the most important aspect of the system.

The data fusion process occurs within the sensor network rather than at a centralized computer and is performed by a specially developed algorithm based on Bayesian statistics. WATS would not use a centralized computer for analysis because researchers found that factors such as latency and available bandwidth tended to create significant bottlenecks. Data processed in the field by the network itself (by transferring small amounts of data between neighboring sensors) is faster and makes the network more scalable.

An important factor in WATS development is ease of deployment, since more sensors both improves the detection rate and reduces false alarms. WATS sensors could be deployed in permanent positions or mounted in vehicles for mobile protection of specific locations. One barrier to the implementation of WATS is the size, weight, energy requirements and cost of currently available wireless sensors. The development of improved sensors is a major component of current research at the Nonproliferation, Arms Control, and International Security (NAI) Directorate at LLNL.

WATS was profiled to the U.S. House of Representatives' Military Research and Development Subcommittee on October 1, 1997 during a hearing on nuclear terrorism and countermeasures. On August 4, 1998 in a subsequent meeting of that subcommittee, Chairman Curt Weldon stated that research funding for WATS had been cut by the Clinton administration to a subsistence level and that the program had been poorly re-organized.

Characteristics

The main characteristics of a WSN include:

- Power consumption constraints for nodes using batteries or energy harvesting. Examples of suppliers are ReVibe Energy and Perpetuum.
- Ability to cope with node failures (resilience).
- Some mobility of nodes (for highly mobile nodes see MWSNs).
- Heterogeneity of nodes.
- Homogeneity of nodes.
- Scalability to large scale of deployment.
- Ability to withstand harsh environmental conditions.
- Ease of use.

- Cross-layer optimization.

Cross-layer is becoming an important studying area for wireless communications. In addition, the traditional layered approach presents three main problems:

- Traditional layered approach cannot share different information among different layers, which leads to each layer not having complete information. The traditional layered approach cannot guarantee the optimization of the entire network.
- The traditional layered approach does not have the ability to adapt to the environmental change.
- Because of the interference between the different users, access conflicts, fading, and the change of environment in the wireless sensor networks, traditional layered approach for wired networks is not applicable to wireless networks.

So the cross-layer can be used to make the optimal modulation to improve the transmission performance, such as data rate, energy efficiency, QoS (Quality of Service), etc. Sensor nodes can be imagined as small computers which are extremely basic in terms of their interfaces and their components. They usually consist of a processing unit with limited computational power and limited memory, sensors or MEMS (including specific conditioning circuitry), a communication device (usually radio transceivers or alternatively optical), and a power source usually in the form of a battery. Other possible inclusions are energy harvesting modules, secondary ASICs, and possibly secondary communication interface (e.g. RS-232 or USB).

The base stations are one or more components of the WSN with much more computational, energy and communication resources. They act as a gateway between sensor nodes and the end user as they typically forward data from the WSN on to a server. Other special components in routing based networks are routers, designed to compute, calculate and distribute the routing tables.

Platforms

Hardware

One major challenge in a WSN is to produce low cost and tiny sensor nodes. There are an increasing number of small companies producing WSN hardware and the commercial situation can be compared to home computing in the 1970s. Many of the nodes are still in the research and development stage, particularly their software. Also inherent to sensor network adoption is the use of very low power methods for radio communication and data acquisition.

In many applications, a WSN communicates with a Local Area Network or Wide Area Network through a gateway. The Gateway acts as a bridge between the WSN and the other network. This enables data to be stored and processed by devices with more resources, for example, in a remotely located server. A wireless wide area network used primarily for low-power devices is known as a Low-Power Wide-Area Network (LPWAN).

Wireless

There are several wireless standards and solutions for sensor node connectivity. Thread and ZigBee can connect sensors operating at 2.4 GHz with a data rate of 250kbit/s. Many use a lower frequency to increase radio range (typically 1 km), for example Z-wave operates at 915 MHz and in the EU 868 MHz has been widely used but these have a lower data rate (typically 50 kb/s). The IEEE 802.15.4 working group provides a standard for low power device connectivity and commonly sensors and smart meters use one of these standards for connectivity. With the emergence of Internet of Things, many other proposals have been made to provide sensor connectivity. LORA is a form of LPWAN which provides long range low power wireless connectivity for devices, which has been used in smart meters. Wi-SUN connects devices at home. NarrowBand IOT and LTE-M can connect up to millions of sensors and devices using cellular technology.

Software

Energy is the scarcest resource of WSN nodes, and it determines the lifetime of WSNs. WSNs may be deployed in large numbers in various environments, including remote and hostile regions, where ad hoc communications are a key component. For this reason, algorithms and protocols need to address the following issues:

- Increased lifespan.
- Robustness and fault tolerance.
- Self-configuration.

Lifetime maximization: Energy/Power Consumption of the sensing device should be minimized and sensor nodes should be energy efficient since their limited energy resource determines their lifetime. To conserve power, wireless sensor nodes normally power off both the radio transmitter and the radio receiver when not in use.

Routing Protocols

Wireless sensor networks are composed of low-energy, small-size, and low-range unattended sensor nodes. Recently, it has been observed that by periodically turning on and off the sensing and communication capabilities of sensor nodes, we can significantly reduce the active time and thus prolong network lifetime. However, this duty cycling may result in high network latency, routing overhead, and neighbor discovery delays due to asynchronous sleep and wake-up scheduling. These limitations call for a countermeasure for duty-cycled wireless sensor networks which should minimize routing information, routing traffic load, and energy consumption. Researchers from Sungkyunkwan University have proposed a lightweight non-increasing delivery-latency interval routing referred as LNDIR. This scheme can discover minimum latency routes at each non-increasing delivery-latency interval instead of each time slot. Simulation experiments demonstrated the validity of this novel approach in minimizing routing information stored at each sensor. Furthermore, this novel routing can also guarantee the minimum delivery latency from each source to the sink. Performance improvements of up to 12-fold and 11-fold are observed in terms of routing traffic load reduction and energy efficiency, respectively, as compared to existing schemes.

Operating Systems

Operating systems for wireless sensor network nodes are typically less complex than general-purpose operating systems. They more strongly resemble embedded systems, for two reasons. First, wireless sensor networks are typically deployed with a particular application in mind, rather than as a general platform. Second, a need for low costs and low power leads most wireless sensor nodes to have low-power microcontrollers ensuring that mechanisms such as virtual memory are either unnecessary or too expensive to implement.

It is therefore possible to use embedded operating systems such as eCos or uC/OS for sensor networks. However, such operating systems are often designed with real-time properties.

TinyOS is perhaps the first operating system specifically designed for wireless sensor networks. TinyOS is based on an event-driven programming model instead of multithreading. TinyOS programs are composed of event handlers and tasks with run-to-completion semantics. When an external event occurs, such as an incoming data packet or a sensor reading, TinyOS signals the appropriate event handler to handle the event. Event handlers can post tasks that are scheduled by the TinyOS kernel some time later.

LiteOS is a newly developed OS for wireless sensor networks, which provides UNIX-like abstraction and support for the C programming language.

Contiki is an OS which uses a simpler programming style in C while providing advances such as 6LoWPAN and Protothreads.

RIOT (operating system) is a more recent real-time OS including similar functionality to Contiki.

PreonVM is an OS for wireless sensor networks, which provides 6LoWPAN based on Contiki and support for the Java programming language.

Online Collaborative Sensor Data Management Platforms

Online collaborative sensor data management platforms are on-line database services that allow sensor owners to register and connect their devices to feed data into an online database for storage and also allow developers to connect to the database and build their own applications based on that data. Examples include Xively and the Wikisensing platform. Such platforms simplify online collaboration between users over diverse data sets ranging from energy and environment data to that collected from transport services. Other services include allowing developers to embed real-time graphs & widgets in websites; analyse and process historical data pulled from the data feeds; send real-time alerts from any datastream to control scripts, devices and environments.

The architecture of the Wikisensing system describes the key components of such systems to include APIs and interfaces for online collaborators, a middleware containing the business logic needed for the sensor data management and processing and a storage model suitable for the efficient storage and retrieval of large volumes of data.

Simulation

At present, agent-based modeling and simulation is the only paradigm which allows the simulation of complex behavior in the environments of wireless sensors (such as flocking). Agent-based simulation of wireless sensor and ad hoc networks is a relatively new paradigm. Agent-based modelling was originally based on social simulation.

Network simulators like Opnet, Tetcos NetSim and NS can be used to simulate a wireless sensor network.

Other Concepts

Security

Infrastructure-less architecture (i.e. no gateways are included, etc.) and inherent requirements (i.e. unattended working environment, etc.) of WSNs might pose several weak points that attract adversaries. Therefore, security is a big concern when WSNs are deployed for special applications such as military and healthcare. Owing to their unique characteristics, traditional security methods of computer networks would be useless (or less effective) for WSNs. Hence, lack of security mechanisms would cause intrusions towards those networks. These intrusions need to be detected and mitigation methods should be applied.

Distributed Sensor Network

If a centralized architecture is used in a sensor network and the central node fails, then the entire network will collapse, however the reliability of the sensor network can be increased by using a distributed control architecture. Distributed control is used in WSNs for the following reasons:

- Sensor nodes are prone to failure.
- For better collection of data.

- To provide nodes with backup in case of failure of the central node.

There is also no centralized body to allocate the resources and they have to be self-organized.

As for the distributed filtering over distributed sensor network. the general setup is to observe the underlying process through a group of sensors organized according to a given network topology, which renders the individual observer estimates the system state based not only on its own measurement but also on its neighbors'.

Data Integration and Sensor Web

The data gathered from wireless sensor networks is usually saved in the form of numerical data in a central base station. Additionally, the Open Geospatial Consortium (OGC) is specifying standards for interoperability interfaces and metadata encodings that enable real time integration of heterogeneous sensor webs into the Internet, allowing any individual to monitor or control wireless sensor networks through a web browser.

In-network Processing

To reduce communication costs some algorithms remove or reduce nodes' redundant sensor information and avoid forwarding data that is of no use. This technique has been used, for instance, for distributed anomaly detection or distributed optimization. As nodes can inspect the data they forward, they can measure averages or directionality for example of readings from other nodes. For example, in sensing and monitoring applications, it is generally the case that neighboring sensor nodes monitoring an environmental feature typically register similar values. This kind of data redundancy due to the spatial correlation between sensor observations inspires techniques for in-network data aggregation and mining. Aggregation reduces the amount of network traffic which helps to reduce energy consumption on sensor nodes. Recently, it has been found those networks gateways also play an important role in improving energy efficiency of sensor nodes by scheduling more resources for the nodes with more critical energy efficiency need and advanced energy efficient scheduling algorithms need to be implemented at network gateways for the improvement of the overall network energy efficiency.

Secure Data Aggregation

This is a form of in-network processing where sensor nodes are assumed to be unsecured with limited available energy, while the base station is assumed to be secure with unlimited available energy. Aggregation complicates the already existing security challenges for wireless sensor networks and requires new security techniques tailored specifically for this scenario. Providing security to aggregate data in wireless sensor networks is known as secure data aggregation in WSN were the first few works discussing techniques for secure data aggregation in wireless sensor networks.

Two main security challenges in secure data aggregation are confidentiality and integrity of data. While encryption is traditionally used to provide end to end confidentiality in wireless sensor network, the aggregators in a secure data aggregation scenario need to decrypt the encrypted data to perform aggregation. This exposes the plaintext at the aggregators, making the data vulnerable to attacks from an adversary. Similarly an aggregator can inject false data into the aggregate and make the base station accept false data. Thus, while data aggregation improves energy efficiency of a network, it complicates the existing security challenges.

Wireless PAN

WPAN works much like a standard personal area network (PAN) except that it uses a wireless communication medium instead of a wired connection. Typically, the devices in WPAN include peripheral and hand-held devices such as PDAs, smart phones and tablet PCs. A WPAN's range depends on the wireless router's capabilities, access point or the device itself, but it is usually restricted to a house or small office. WPAN can be created using Wi-Fi, Bluetooth, infrared, Z-wave or any similar wireless technologies. In some cases, one of the Internet enabled/powered devices acts as an access point and provides network and Internet access to other devices.

For example, a laptop can be connected to the Internet wirelessly by creating a Bluetooth WPAN with a cell phone. The General Packet Radio Service (GPRS) Internet connectivity of the cell phone can be shared with the laptop and all data packets to and from the laptop are sent over the Bluetooth-powered WPAN.

Wireless PAN Systems generally apply to individual users, and some offer support for multiple users.

Home and Small Office

Many different system configurations of wireless PANs exist in the home and small office.

Synchronization

One of the most common uses of wireless PANs is PDA and cell phone synchronization with a laptop or PC. The interconnection of components for this type of system. When the user presses a sync button on the handheld device, the radio NIC within the handheld device sends the corresponding data to the radio NIC in the laptop or PC. Likewise, the laptop or PC will send data to the handheld device. In most cases, the wireless connection extends the serial RS-232 port wirelessly to the handheld device.

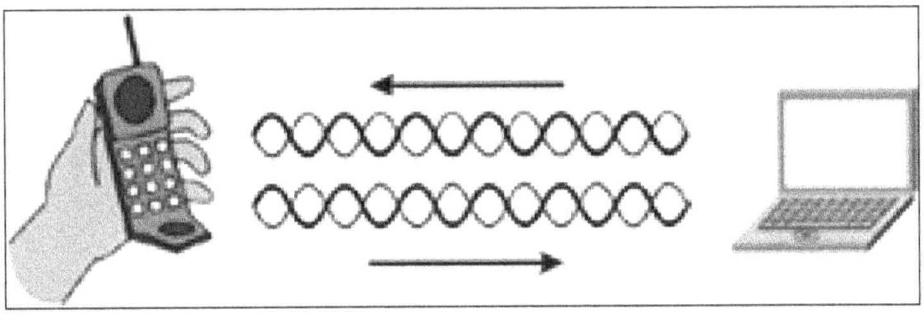

Synchronization transfers information in both directions between two devices.

Streaming Multimedia

A large number of wireless PAN applications involve streaming audio and video. For example, a user can easily listen to streaming MP3 files stored on an MP3 player. Many PDAs have the capability of playing MP3 audio files by installing one of the popular media players, such as the RealOne media player from RealNetworks, Inc. With a wireless PAN, the user doesn't need to carry around the MP3 player and mess with wires or stay within the same area to listen to music. A similar configuration involves the use of a wireless audio earpiece and microphone for a hands-free operation of a cell phone. A drawback to this approach, however, is that batteries will not last as long when using the wireless connections.

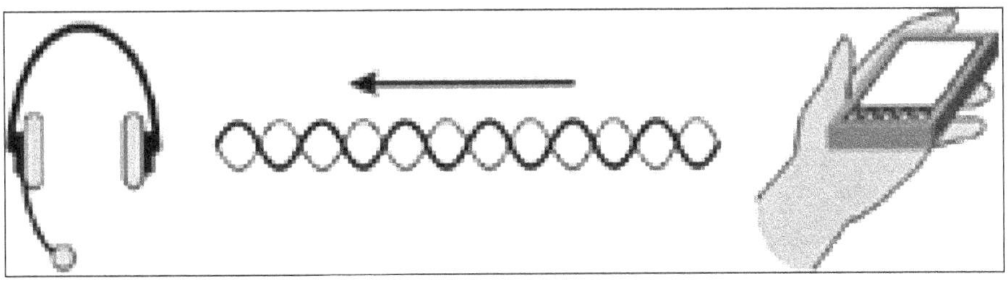

Wireless pans permit easy use of headphones.

Another benefit of wireless PANs in streaming applications is flexible connectivity between video cameras and a server. A homeowner could, for example, place web cams in strategic places for security monitoring purposes. A hidden camera aimed at the front door area allows the homeowner to screen visitors before opening the door. The use of wireless, in this case, simplifies the installation because it eliminates the need to run wires to the camera. Electrical current or batteries are still necessary to power the camera, of course, but electrical outlets are available throughout a home.

Control

Wireless PANs eliminate wires for computer peripherals, such as a wireless mouse, keyboard, and telephone connection, making it easier to move and set up PCs. A user, for example, can use a full-sized keyboard wirelessly with a laptop or PDA. In addition, wireless PANs reduce the tangle of cables surrounding a desktop computer. Reliability is higher because of less cable breakage and less risk of someone inadvertently kicking a cable loose.

Printing

Wireless connections between your PC and printer are made possible within the same room through a wireless PAN connection. Printer cables are often too short, and you're stuck setting the printer in a less than ideal location. The wireless PAN connection allows the movement of the printer to a better location.

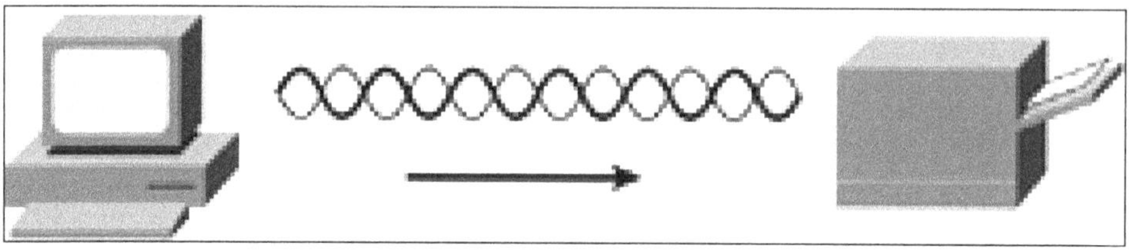

Printing is easy with a wireless PAN.

Internet Connections

A user can access e-mail and browse the web from anywhere within the room with a wireless PAN interface to the Internet. Instead of sitting at a desk, for example, a person can relax in a lounge chair or couch. This freedom makes computing much more enjoyable.

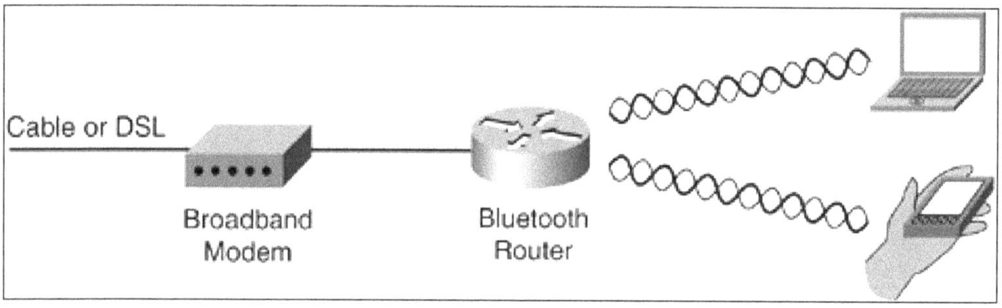

A wireless PAN router allows connectivity to the internet.

Enterprise

The use of wireless PANs in enterprise is common; however, the applications and system configurations are similar to homes and small offices. Employees use wireless PANs to synchronize PDAs with desktop computers and take advantage of wireless peripherals. Instead of using a wireless PAN router for connecting users to the Internet however, an enterprise makes use of wireless LANs for Internet connectivity. Enterprises span too large of an area to make wireless PANs practical because of the rather large number of required base stations.

Wireless PAN Technologies

Wireless PAN Technologies utilize both radio frequencies and infrared light, depending on the application.

802.15

The IEEE 802.15 standards working group focuses on the development of standards for wireless PANs and coordinates with other standards, such as 802.11 wireless LANs.

The 802.15 standards working group contains the following elements:

- 802.15.1: This working group, Task Group 1, defines a wireless PAN standard based on Bluetooth v1.1 specifications, which uses frequency hopping spread spectrum (FHSS) and operates at up to 1 Mbps. The 802.15 group published 802.11.1 in June of 2002, and it is meant to serve as a resource for developers of Bluetooth devices.
- 802.15.2: The group responsible for this standard, Task Group 2, is defining recommended practices to facilitate the coexistence of 802.15 and 802.11 networks. An issue is that both networks operate in the same 2.4 GHz frequency band, making coordination between operations necessary. The group is quantifying the interference and proposing methods to counter the interference.
- 802.15.3: This is Task Group 3, which is drafting a new standard for higher-rate wireless PANs. Data rates include 11, 22, 33, 44, and 55 Mbps. Combined with these higher data rates, quality of service (QoS) mechanisms make this standard good for satisfying needs for multimedia applications. This group is also focusing on lower cost and power requirements. A draft of the 802.15.3 standard is now available for purchase.
- 802.15.4: This group, Task Group 4, is investigating the definition of a standard with low data rates that leads to extremely low-power consumption for small devices where it's not practical to change batteries within months or years. For example, sensors, smart badges, and home automation systems are candidates for this technology. Data rates include 20, 40, and 250 kbps. A draft of the 802.15.4 standard is now available for purchase.

Bluetooth

The introduction of Bluetooth in 1998 was the result of several companies, including Ericsson, IBM, Intel, Nokia, and Toshiba, working together to create a solution for wireless access among computing devices. Bluetooth, which is a specification and not a standard, is ideal for small devices with short-range, low-power, and inexpensive radio links. This makes Bluetooth a good solution for connecting small devices within range of a person in a small working area. That's why the 802.15 chose Bluetooth as the basis of the 802.15.1 standard.

Basic Features

The Bluetooth Special Interest Group (SIG) published the initial version of the specification in mid-1999. There have been updates since then, but the technical attributes are essentially the same. Bluetooth transceivers operate at up to 1 Mbps data rate in the 2.4GHz band, using FHSS technology. It constantly hops over the entire spectrum at a rate of 1,600 hops per second, which is much faster than the 802.11 version of frequency hopping.

Low-power Bluetooth devices have a range of 30 feet. High-power Bluetooth devices, however, can reach distances of around 300 feet. The high-power mode, though, is rare.

Bluetooth modules have relatively small form factors. Typical measurements are 10.2 x 14 x 1.6 millimeters, which is small enough to fit in a variety of user devices.

Bluetooth enables automatic connection among Bluetooth devices that fall within range of each other, but a user has the ability to accept and disallow connections with specific users. Users, however, should always be aware of whether their Bluetooth connection is enabled. To ensure security, disable the Bluetooth connection. Encryption is also part of the specification.

Bluetooth has characteristics similar to wireless LANs. Through the use of the high-power version of Bluetooth, manufacturers can develop Bluetooth access points and routers with a similar range as 802.11 networks. The current Bluetooth products, however, are mostly low power and focus on wireless PAN functions. In addition, it would be difficult for any Bluetooth wireless LAN products to gain a strong foothold in the market because 802.11 products already have widespread adoption.

The place where Bluetooth falls behind 802.11 is performance and range. 802.11 components can reach data rates of up to 54 Mbps, while Bluetooth lags way behind at around 1 Mbps. This might be good enough for most cable replacement applications, such as an interface between headphones and a PDA but higher performance is necessary when surfing the web through a broadband connection or participating on a corporate network. Also, the range of 802.11 is typically 300 feet inside offices, which is much greater than Bluetooth. Bluetooth would require many access points to fully cover larger areas.

As a result, it's highly unlikely that Bluetooth products will win over 802.11. This is certainly apparent because electronics stores primarily sell 802.11 (Wi-Fi) solutions for wireless LAN applications, not Bluetooth.

It's possible that 802.11 wireless LANs could have a big impact on the sale of Bluetooth devices, mostly because 802.11 meets or exceeds nearly all of the characteristics of Bluetooth. Because widespread adoption of Bluetooth is still lacking, there's time for 802.11 vendors to get their foot in the door with manufacturers needing support for wireless PANs.

Some modifications would need to be made, however. The size of 802.11 components needs to be smaller, but that is becoming more of a reality as semiconductor companies strive for miniaturization of their 802.11 chipsets. These smaller components require less power, making them more competitive for devices, such as mobile phones, that have smaller batteries. With the 802.15 group defining standards for wireless PANs based on Bluetooth and the 802.11 group focusing on wireless LANs it's likely that both Bluetooth and 802.11 will continue to coexist and complement each other.

Minimizing Bluetooth Interference

As more wireless products become available, you need to carefully manage potential frequency interference. Tests have shown significant interference between Bluetooth and other systems operating in the 2.4 GHz band, such as 802.11 wireless LANs. A critical problem is that Bluetooth and 802.11b neither understand each other nor follow the same rules. A Bluetooth radio might haphazardly begin transmitting data while an 802.11 station is sending a frame. This results in a collision, which forces the 802.11 station to retransmit the frame. This lack of coordination is the basis for radio frequency (RF) interference between Bluetooth and 802.11.

Because of the potential for collisions, 802.11 and Bluetooth networks suffer from lower performance. An 802.11 station automatically lowers its data rate and retransmits a frame when collisions occur. Consequently, the 802.11 protocol introduces delays in the presence of Bluetooth interference.

The full impact of RF interference depends on the utilization and proximity of Bluetooth devices. Interference occurs only when both Bluetooth and 802.11b devices transmit at the same time. Users might have Bluetooth devices in their PDAs or laptops, but no interference will exist if their applications are not using the Bluetooth radio to send data.

Some Bluetooth applications, such as printing from a laptop or synchronizing a PDA to a desktop, utilize the radio for a short period of time. In this case, the Bluetooth devices are not active long enough to noticeably degrade the performance of an 802.11 network. For example, a user might synchronize her PDA to her desktop when arriving at work in the morning. Other than that, their Bluetooth radio might be inactive and not cause interference the rest of the day.

The biggest impact is when a company implements a large-scale Bluetooth network, such as one that enables mobility for doctors and nurses using PDAs throughout a hospital. If the Bluetooth network is widespread and under moderate-to-high levels of utilization, the Bluetooth system will probably offer a substantial number of collisions with an 802.11 network residing in the same area. In this case, Bluetooth and 802.11 would have difficulties coexisting, and performance would likely suffer.

In addition to utilization, the proximity of the Bluetooth devices to 802.11 radio NICs and access points has a tremendous effect on the degree of interference. The transmit power of Bluetooth devices is generally lower than 802.11 wireless LANs. Therefore, an 802.11 station must be relatively close (within 10 feet or so) of a transmitting Bluetooth device before significant interference can occur.

A typical application fitting this scenario is a laptop user utilizing Bluetooth to support connections to a PDA and printer and 802.11 to access the Internet and corporate servers. The potential for interference in this situation is enormous, especially when the user is operating within outer limits of the coverage area of the 802.11 network. The signal from the Bluetooth device will likely drown out the weaker 802.11 signal because of the distance of the access point.

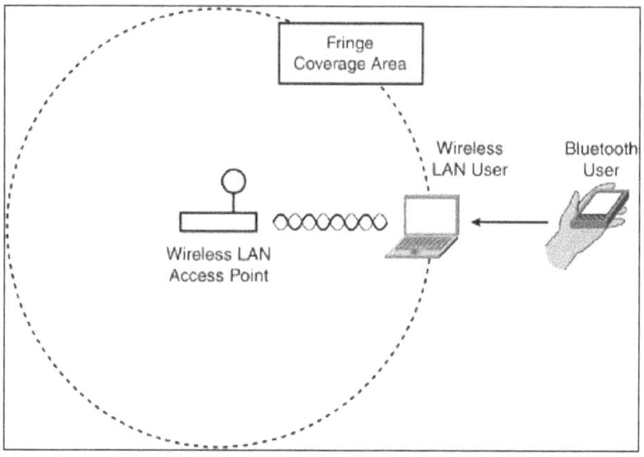

Rf interference can occur between Bluetooth and 802.11 wireless LAN devices.

Here are some tips on how to avoid interference from Bluetooth devices:

- Manage the use of RF devices: One way to reduce the potential for interference is to regulate the types of RF devices within your home or office. In other words, establish your own private regulatory body for managing unlicensed RF devices. The extreme measure would be to completely ban the use of Bluetooth; however, that is not practical or even possible in all cases. For example, you can't feasibly prohibit the use of Bluetooth in public areas of large offices. For private applications, you could set company policies to limit the use of Bluetooth to specific applications, such as synchronizing PDAs to desktops.
- Ensure adequate 802.11 coverage: Strong, healthy 802.11 signals throughout the coverage areas reduce the impact of the Bluetooth signals. If wireless LAN transmissions become too weak, the interfering Bluetooth signals will be more troublesome. Perform a thorough RF site survey, and determine the appropriate location for access points.
- Move to the 5 GHz band: If none of the preceding steps solve the problem, consider using a 5 GHz wireless LAN such as 802.11a. You can completely avoid RF interference in this band? at least for the foreseeable future.

IrDA

Bluetooth's primary competitor is Infrared Data Association (IrDA), which has been defining and publishing since 1993. The IrDA has a charter to create an interoperable, low-cost, low-power, serial data communications standard for short-range applications. IrDA has been around for much longer than Bluetooth. In fact, many laptops and cell phones have been coming equipped with an IrDA interface for years.

Basic Features

The basis for IrDA is infrared light, which doesn't go through walls and other obstacles. This strictly limits the range of IrDA devices to within an obstacle-free room. This makes IrDA useful only for point-to-point applications, such as synchronizing PDAs to PCs. An advantage of IrDA, however, is that there's no worry about RF interference.

The IrDA data standard, which is best for devices such as an MP3 player needing to stream information, offers up to 4 Mbps data rates. This version of the standard has up to 3 feet (1 meter range), but low-power versions significantly conserve battery power and reduce operation to approximately 8 inches (20 centimeters).

To effectively support wireless computer peripherals, such as a keyboard or mouse, the IrDA control version of the standard reduces data rates to 75 kbps. In addition, the host computer can communicate with up to eight peripherals simultaneously.

Wireless LAN

A wireless LAN (WLAN) is a wireless computer network that links two or more devices using wireless communication to form a local area network (LAN) within a limited area such as a home, school, computer laboratory, campus, or office building. This gives users the ability to move around within the area and remain connected to the network. Through a gateway, a WLAN can also provide a connection to the wider Internet.

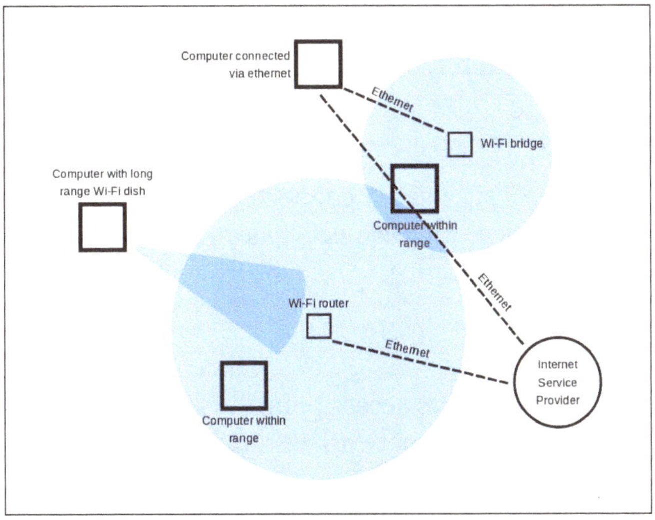

An example of a WiFi network.

Most modern WLANs are based on IEEE 802.11 standards and are marketed under the Wi-Fi brand name.

Wireless LANs have become popular for use in the home, due to their ease of installation and use. They are also popular in commercial properties that offer wireless access to their employees and customers.

Architecture

Stations

All components that can connect into a wireless medium in a network are referred to as stations (STA). All stations are equipped with wireless network interface controllers (WNICs). Wireless stations fall into two categories: wireless access points, and clients. Access points (APs), normally wireless routers, are base stations for the wireless network. They transmit and receive radio frequencies for wireless enabled devices to communicate with. Wireless clients can be mobile devices such as laptops, personal digital assistants, IP phones and other smartphones, or non-portable devices such as desktop computers, printers, and workstations that are equipped with a wireless network interface.

Basic Service Set

The basic service set (BSS) is a set of all stations that can communicate with each other at PHY layer. Every BSS has an identification (ID) called the BSSID, which is the MAC address of the access point servicing the BSS.

There are two types of BSS: Independent BSS (also referred to as IBSS), and infrastructure BSS. An independent BSS (IBSS) is an ad hoc network that contains no access points, which means they cannot connect to any other basic service set.

Independent Basic Service Set

An IBSS is a set of STAs configured in ad hoc (peer-to-peer) mode.

Extended Service Set

An extended service set (ESS) is a set of connected BSSs. Access points in an ESS are connected by a distribution system. Each ESS has an ID called the SSID which is a 32-byte (maximum) character string.

Distribution System

A distribution system (DS) connects access points in an extended service set. The concept of a DS can be used to increase network coverage through roaming between cells.

DS can be wired or wireless. Current wireless distribution systems are mostly based on WDS or MESH protocols, though other systems are in use.

Types of Wireless LANs

The IEEE 802.11 has two basic modes of operation: Infrastructure and ad hoc mode. In ad hoc mode, mobile units transmit directly peer-to-peer. In infrastructure mode, mobile units communicate through an access point that serves as a bridge to other networks (such as Internet or LAN).

Since wireless communication uses a more open medium for communication in comparison to wired LANs, the 802.11 designers also included encryption mechanisms: Wired Equivalent Privacy (WEP, now insecure), Wi-Fi Protected Access (WPA, WPA2, WPA3), to secure wireless computer networks. Many access points will also offer Wi-Fi Protected Setup, a quick (but now insecure) method of joining a new device to an encrypted network.

Infrastructure

Most Wi-Fi networks are deployed in infrastructure mode.

In infrastructure mode, a base station acts as a wireless access point hub, and nodes communicate through the hub. The hub usually, but not always, has a wired or fiber network connection, and may have permanent wireless connections to other nodes.

Wireless access points are usually fixed, and provide service to their client nodes within range.

Wireless clients, such as laptops and smartphones, connect to the access point to join the network.

Sometimes a network will have a multiple access points, with the same 'SSID' and security arrangement. In that case connecting to any access point on that network joins the client to the network. In that case, the client software will try to choose the access point to try to give the best service, such as the access point with the strongest signal.

Peer-to-peer

An ad hoc network (not the same as a WiFi Direct network) is a network where stations communicate only peer to peer (P2P). There is no base and no one gives permission to talk. This is accomplished using the Independent Basic Service Set (IBSS).

A WiFi Direct network is another type of network where stations communicate peer to peer.

In a Wi-Fi P2P group, the group owner operates as an access point and all other devices are clients. There are two main methods to establish a group owner in the Wi-Fi Direct group. In one approach, the user sets up a P2P group owner manually. This method is also known as Autonomous Group Owner (autonomous GO). In the second method, also called negotiation-based group creation, two devices compete based on the group owner intent value. The device with higher intent value becomes a group owner and the second device becomes a client. Group owner intent value can depend on whether the wireless device performs a cross-connection between an infrastructure WLAN service and a P2P group, remaining power in the wireless device, whether the wireless device is already a group owner in another group and/or a received signal strength of the first wireless device.

A peer-to-peer network allows wireless devices to directly communicate with each other. Wireless devices within range of each other can discover and communicate directly without involving central access points. This method is typically used by two computers so that they can connect to each other to form a network. This can basically occur in devices within a closed range.

If a signal strength meter is used in this situation, it may not read the strength accurately and can be misleading, because it registers the strength of the strongest signal, which may be the closest computer.

IEEE 802.11 defines the physical layer (PHY) and MAC (Media Access Control) layers based on CSMA/CA (Carrier Sense Multiple Access with Collision Avoidance). This is in contrast to Ethernet which uses CSMA-CD (Carrier Sense Multiple Access with Collision Detection). The 802.11 specification includes provisions designed to minimize collisions, because two mobile units may both be in range of a common access point, but out of range of each other.

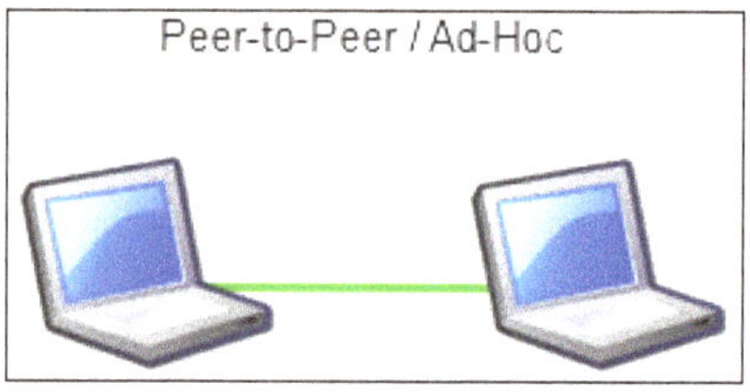

Peer-to-Peer or ad hoc wireless LAN.

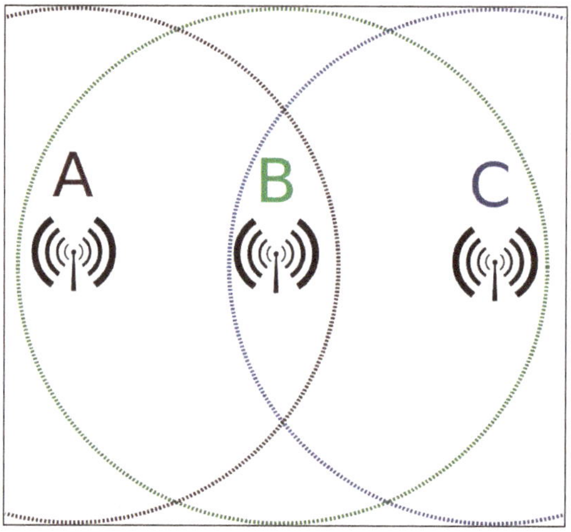

Hidden node problem: Devices A and C are both communicating with B, but are unaware of each other.

Bridge

A bridge can be used to connect networks, typically of different types. A wireless Ethernet bridge allows the connection of devices on a wired Ethernet network to a wireless network. The bridge acts as the connection point to the Wireless LAN.

Wireless Distribution System

A wireless distribution system (WDS) enables the wireless interconnection of access points in an IEEE 802.11 network. It allows a wireless network to be expanded using multiple access points without the need for a wired backbone to link them, as is traditionally required. The notable advantage of a WDS over other solutions is that it preserves the MAC addresses of client packets across links between access points.

An access point can be either a main, relay or remote base station. A main base station is typically connected to the wired Ethernet. A relay base station relays data between remote base stations, wireless clients or other relay stations to either a main or another relay base station. A remote base station accepts connections from wireless clients and passes them to relay or main stations. Connections between clients are made using MAC addresses rather than by specifying IP assignments.

All base stations in a WDS must be configured to use the same radio channel, and share WEP keys or WPA keys if they are used. They can be configured to different service set identifiers. WDS also requires that every base station be configured to forward to others in the system.

WDS capability may also be referred to as repeater mode because it appears to bridge and accept wireless clients at the same time (unlike traditional bridging). Throughput in this method is halved for all clients connected wirelessly.

When it is difficult to connect all of the access points in a network by wires, it is also possible to put up access points as repeaters.

Roaming

There are two definitions for wireless LAN roaming:

- Internal roaming: The Mobile Station (MS) moves from one access point (AP) to another AP within a home network if the signal strength is too weak. An authentication server (RADIUS) performs the re-authentication of MS via 802.1x (e.g. with PEAP). The billing of QoS is in the home network. A Mobile Station roaming from one access point to another often interrupts the flow of data among the Mobile Station and an application connected to the network. The Mobile Station, for instance, periodically monitors the presence of alternative access points (ones that will provide a better connection). At some point, based on proprietary mechanisms, the Mobile Station decides to re-associate with an access point having a stronger wireless signal. The Mobile Station, however, may lose a connection with an access point before associating with another access point. In order to provide reliable connections with applications, the Mobile Station must generally include software that provides session persistence.

- External roaming: The MS (client) moves into a WLAN of another Wireless Internet Service Provider (WISP) and takes their services (Hotspot). The user can use a foreign network independently from their home network, provided that the foreign network allows visiting users on their network. There must be special authentication and billing systems for mobile services in a foreign network.

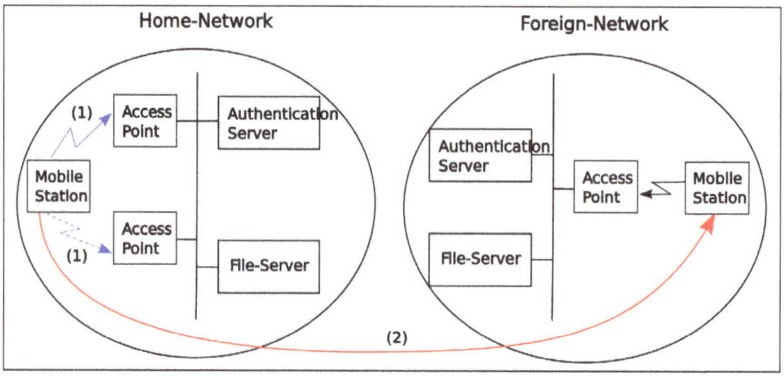

Applications

Wireless LANs have a great deal of applications. Modern implementations of WLANs range from small in-home networks to large, campus-sized ones to completely mobile networks on airplanes and trains.

Users can access the Internet from WLAN hotspots in restaurants, hotels, and now with portable devices that connect to 3G or 4G networks. Oftentimes these types of public access points require no registration or password to join the network. Others can be accessed once registration has occurred and/or a fee is paid.

Existing Wireless LAN infrastructures can also be used to work as indoor positioning systems with no modification to the existing hardware.

Wireless Ad Hoc Network

A wireless ad hoc network (WANET) or Mobile ad hoc network (MANET) is a decentralised type of wireless network. The network is ad hoc because it does not rely on a pre-existing infrastructure, such as routers in wired networks or access points in managed (infrastructure) wireless networks. Instead, each node participates in routing by forwarding data for other nodes, so the determination of which nodes forward data is made dynamically on the basis of network connectivity and the routing algorithm in use.

In the Windows operating system, ad-hoc is a communication mode (setting) that allows computers to directly communicate with each other without a router. Wireless mobile ad hoc networks are self-configuring, dynamic networks in which nodes are free to move.

Such wireless networks lack the complexities of infrastructure setup and administration, enabling devices to create and join networks "on the fly" – anywhere, anytime.

Each device in a MANET is free to move independently in any direction, and will therefore change its links to other devices frequently. Each must forward traffic unrelated to its own use, and therefore be a router. The primary challenge in building a MANET is equipping each device to continuously maintain the information required to properly route traffic. Such networks may operate by themselves or may be connected to the larger Internet. They may contain one or multiple and different transceivers between nodes. This results in a highly dynamic, autonomous topology.

MANETs usually have a routable networking environment on top of a Link Layer ad hoc network. MANETs consist of a peer-to-peer, self-forming, self-healing network. MANETs circa 2000–2015 typically communicate at radio frequencies (30 MHz – 5 GHz).

Applications

The decentralized nature of wireless ad-hoc networks makes them suitable for a variety of applications where central nodes can't be relied on and may improve the scalability of networks compared to wireless managed networks, though theoretical and practical limits to the overall capacity of such networks have been identified. Minimal configuration and quick deployment make ad hoc networks suitable for emergency situations like natural disasters or military conflicts. The presence of dynamic and adaptive routing protocols enables ad hoc networks to be formed quickly. Wireless ad-hoc networks can be further classified by their applications:

Mobile Ad Hoc Networks (MANETs)

A mobile ad hoc network (MANET) is a continuously self-configuring, self-organizing, infrastructure-less network of mobile devices connected without wires. It is sometimes known as "on-the-fly" networks or "spontaneous networks".

Vehicular Ad Hoc Networks (VANETs)

VANETs are used for communication between vehicles and roadside equipment. Intelligent vehicular ad hoc networks (InVANETs) are a kind of artificial intelligence that helps vehicles to behave in intelligent manners during vehicle-to-vehicle collisions, accidents. Vehicles are using radio waves to communicate with each other, creating communication networks instantly on-the-fly while vehicles move along roads.

Smartphone Ad Hoc Networks (SPANs)

A SPAN leverages existing hardware (primarily Wi-Fi and Bluetooth) and software (protocols) in commercially available smartphones to create peer-to-peer networks without relying on cellular carrier networks, wireless access points, or traditional network infrastructure. SPANs differ from traditional hub and spoke networks, such as Wi-Fi Direct, in that they support multi-hop relays and there is no notion of a group leader so peers can join and leave at will without destroying the network. Most recently, Apple's iPhone with version 8.4 iOS and higher have been enabled with multi-peer ad hoc mesh networking capability, in iPhones, allowing millions of smart phones to create ad hoc networks without relying on cellular communications. It has been claimed that this is going to "change the world".

iMANETs

Internet-based mobile ad-hoc networks (iMANETs) is a type of wireless ad hoc network that supports Internet protocols such as TCP/UDP and IP. The network uses a network-layer routing protocol to link mobile nodes and establish routes distributed and automatically.

Wireless Mesh Networks

Mesh networks take their name from the topology of the resultant network. In a fully connected mesh, each node is connected to every other node, forming a "mesh". A partial mesh, by contrast, has a topology in which some nodes are not connected to others, although this term is seldom in use. Wireless ad hoc networks can take the form of mesh networks or others. A wireless ad hoc network does not have fixed topology, and its connectivity among nodes is totally dependent on the behavior of the devices, their mobility patterns, distance with each other, etc. Hence, wireless mesh networks are a particular type of wireless ad hoc networks, with special emphasis on the resultant network topology. While some wireless mesh networks (particularly those within a home) have relatively infrequent mobility and the thus infrequent link breaks, other more mobile mesh networks require frequent routing adjustments to account for lost links. Google Home, Google Wi-Fi, and Google OnHub all support Wi-Fi mesh (i.e., Wi-Fi ad hoc) networking. Apple's AirPort allows the formation of wireless mesh networks at home, connecting various Wi-Fi devices together and providing good wireless coverage and connectivity at home.

Army Tactical MANETs

Military or tactical MANETs are used by military units with emphasis on data rate, real-time requirement, fast re-routing during mobility, data security, radio range, and integration with existing systems. Common radio waveforms include the US Army's JTRS SRW and Persistent System's WaveRelay. Ad hoc mobile communications come in well to fulfill this need, especially its infrastructureless nature, fast deployment and operation. Military MANETs are used by military units with emphasis on rapid deployment, infrastructureless, all-wireless networks (no fixed radio towers), robustness (link breaks are no problem), security, range, and instant operation. MANETs can be used in army "hopping" mines, in platoons where soldiers communicate in foreign terrains, giving them superiority in the battlefield. Tactical MANETs can be formed automatically during the mission and the network "disappears" when the mission is over or decommissioned. It is sometimes called "on-the-fly" wireless tactical network.

Air Force UAV Ad Hoc Networks

Flying ad hoc networks (FANETs) are composed of unmanned aerial vehicles, allowing great mobility and providing connectivity to remote areas.

Unmanned aerial vehicle is an aircraft with no pilot on board. UAVs can be remotely controlled (i.e., flown by a pilot at a ground control station) or can fly autonomously based on pre-programmed flight plans. Civilian usage of UAV include modeling 3D terrains, package delivery (Amazon), etc.

UAVs have also been used by US Air Force for data collection and situation sensing, without risking the pilot in a foreign unfriendly environment. With wireless ad hoc network technology embedded into the UAVs, multiple UAVs can communicate with each other and work as a team, collaboratively to complete a task and mission. If a UAV is destroyed by an enemy, its data can be quickly offloaded wirelessly to other neighboring UAVs. The UAV ad hoc communication network is also sometimes referred to UAV instant sky network.

Navy Ad Hoc Networks

Navy ships traditionally use satellite communications and other maritime radios to communicate with each other or with ground station back on land. However, such communications are restricted by delays and limited bandwidth. Wireless ad hoc networks enable ship-area-networks to be formed while at sea, enabling high speed wireless communications among ships, enhancing their sharing of imaging and multimedia data, and better co-ordination in battlefield operations. Some defense companies (such as Rockwell Collins and Rohde & Schwartz) have produced products that enhance ship-to-ship and ship-to-shore communications.

Wireless Sensor Networks

Sensors are useful devices that collect information related to a specific parameter, such as noise, temperature, humidity, pressure, etc. Sensors are increasingly connected via wireless to allow large scale collection of sensor data. With a large sample of sensor data, analytics processing can be used to make sense out of these data. The connectivity of wireless sensor networks rely on the principles behind wireless ad hoc networks, since sensors can now be deploy without any fixed radio towers, and they can now form networks on-the-fly. “Smart Dust” was one of the early projects done at U C Berkeley, where tiny radios were used to interconnect smart dust. More recently, mobile wireless sensor networks (MWSNs) have also become an area of academic interest.

Ad Hoc Home Smart Lighting

ZigBee is a low power form of wireless ad hoc networks that is now finding their way in home automation. Its low power consumption, robustness and extended range inherent in mesh networking can deliver several advantages for smart lighting in homes and in offices. The control includes adjusting dimmable lights, color lights, and color or scene. The networks allow a set or subset of lights to be controlled over a smart phone or via a computer. The home automation market is tipped to exceed $16 billion by 2019.

Ad Hoc Street Light Networks

Wireless ad hoc smart street light networks are beginning to evolve. The concept is to use wireless control of city street lights for better energy efficiency, as part of a smart city architectural feature. Multiple street lights form a wireless ad hoc network. A single gateway device can control up to 500 street lights. Using the gateway device, one can turn individual lights On, Off or dim them, as well as find out which individual light is faulty and in need of maintenance.

Ad Hoc Network of Robots

Robots are mechanical systems that drive automation and perform chores that would seem difficult for man. Efforts have been made to co-ordinate and control a group of robots to undertake collaborative work to complete a task. Centralized control is often based on a “star” approach, where robots take turns to talk to the controller station. However, with wireless ad hoc networks, robots can form a communication network on-the-fly, i.e., robots can now “talk” to each other and collaborate in a distributed fashion. With a network of robots, the robots can communicate among themselves, share local information, and distributively decide how to resolve a task in the most effective and efficient way.

Disaster Rescue Ad Hoc Network

Another civilian use of wireless ad hoc network is public safety. At times of disasters (floods, storms, earthquakes, fires, etc), a quick and instant wireless communication network is necessary. Especially at times of earthquakes when radio towers had collapsed or were destroyed, wireless ad hoc networks can be formed independently. Firemen and rescue workers can use ad hoc networks to communicate and rescue those injured. Commercial radios with such capability are available on the market.

Hospital Ad Hoc Network

Wireless ad hoc networks allow sensors, videos, instruments, and other devices to be deployed and interconnected wirelessly for clinic and hospital patient monitoring, doctor and nurses alert notification, and also making senses of such data quickly at fusion points, so that lives can be saved.

Data Monitoring and Mining

MANETS can be used for facilitating the collection of sensor data for data mining for a variety of applications such as air pollution monitoring and different types of architectures can be used for such applications. A key characteristic of such applications is that nearby sensor nodes monitoring an environmental feature typically register similar values. This kind of data redundancy due to the spatial correlation between sensor observations inspires the techniques for in-network data aggregation and mining. By measuring the spatial correlation between data sampled by different sensors, a wide class of specialized algorithms can be developed to develop more efficient spatial data mining algorithms as well as more efficient routing strategies. Also, researchers have developed performance models for MANET to apply queueing theory.

Challenges

Advantages for Users

The obvious appeal of MANETs is that the network is decentralised and nodes/devices are mobile, that is to say there is no fixed infrastructure which provides the possibility for numerous applications in different areas such as environmental monitoring, disaster relief and military communications. Since the early 2000s interest in MANETs has greatly increased which, in part, is due to the fact mobility can improve network capacity, shown by Grossglauser and Tse along with the introduction of new technologies.

One main advantage to a decentralised network is that they are typically more robust than centralised networks due to the multi-hop fashion in which information is relayed. For example, in the cellular network setting, a drop in coverage occurs if a base station stops working; however the chance of a single point of failure in a MANET is reduced significantly since the data can take multiple paths. Since the MANET architecture evolves with time it has the potential to resolve issues such as isolation/disconnection from the network. Further advantages of MANETS over networks with a fixed topology include flexibility (an ad hoc network can be created anywhere with mobile devices), scalability (you can easily add more nodes to the network) and lower administration costs (no need to build an infrastructure first).

- Highly performing network.
- No expensive infrastructure must be installed.

- Quick distribution of information around sender.
- No single point of failure.
- Multi hop.
- Scalability.

Implementation Difficulties

With a time evolving network it is clear we should expect variations in network performance due to no fixed architecture (no fixed connections). Furthermore, since network topology determines interference and thus connectivity, the mobility pattern of devices within the network will impact on network performance, possibly resulting in data having to be resent a lot of times (increased delay) and finally allocation of network resources such as power remains unclear. Finally, finding a model that accurately represents human mobility whilst remaining mathematically tractable remains an open problem due to the large range of factors that influence it. Some typical models used include the random walk, random waypoint and levy flight models.

- All network entities may be mobile, so a very dynamic topology is needed.
- Network functions must have a high degree of adaptability.
- There are no central entities, so operations must be managed in a completely distributed manner.
- Battery constraints.

Side Effects

- Use of unlicensed frequency spectrum, contributing to radio spectrum pollution.

Radios for Ad Hoc

Wireless ad hoc networks can operate over different types of radios. They can be UHF (300 – 3000 MHz), SHF (3 – 30 GHz), and EHF (30 – 300 GHz). Wi-Fi ad hoc uses the unlicensed ISM 2.4 GHz radios. They can also be used on 5.8 GHz radios.

Next generation Wi-Fi known as 802.11ax provides low delay, high capacity (up to 10Gbit/s) and low packet loss rate, offering 12 streams – 8 streams at 5 GHz and 4 streams at 2.4 GHz. IEEE 802.11ax uses 8x8 MU-MIMO, OFDMA, and 80 MHz channels. Hence, 802.11ax has the ability to form high capacity Wi-Fi ad hoc networks.

At 60 GHz, there is another form of Wi-Fi known as WiGi – wireless gigabit. This has the ability to offer up to 7Gbit/s throughput. Currently, WiGi is targeted to work with 5G cellular networks.

The higher the frequency, such as those of 300 GHz, absorption of the signal will be more predominant. Army tactical radios usually employ a variety of UHF and SHF radios, including those of VHF to provide a variety of communication modes. At the 800, 900, 1200, 1800 MHz range, cellular radios are predominant. Some cellular radios use ad hoc communications to extend cellular range to areas and devices not reachable by the cellular base station.

Protocol Stack

The challenges affecting MANETs span from various layers of the OSI protocol stack. The media access layer (MAC) has to be improved to resolve collisions and hidden terminal problems. The network layer routing protocol has to be improved to resolve dynamically changing network topologies and broken routes. The transport layer protocol has to be improved to handle lost or broken connections. The session layer protocol has to deal with discovery of servers and services.

A major limitation with mobile nodes is that they have high mobility, causing links to be frequently broken and reestablished. Moreover, the bandwidth of a wireless channel is also limited, and nodes operate on limited battery power, which will eventually be exhausted. These factors make the design of a mobile ad hoc network challenging.

The cross-layer design deviates from the traditional network design approach in which each layer of the stack would be made to operate independently. The modified transmission power will help that node to dynamically vary its propagation range at the physical layer. This is because the propagation distance is always directly proportional to transmission power. This information is passed from the physical layer to the network layer so that it can take optimal decisions in routing protocols. A major advantage of this protocol is that it allows access of information between physical layer and top layers (MAC and network layer).

Some elements of the software stack were developed to allow code updates in situ, i.e., with the nodes embedded in their physical environment and without needing to bring the nodes back into the lab facility. Such software updating relied on epidemic mode of dissemination of information and had to be done both efficiently (few network transmissions) and fast.

Routing

Routing in wireless ad hoc networks or MANETs generally falls into three categories, namely: (a) proactive routing, (b) reacting routing, and (c) hybrid routing.

Proactive Routing

This type of protocols maintains fresh lists of destinations and their routes by periodically distributing routing tables throughout the network. The main disadvantages of such algorithms are:

- Respective amount of data for maintenance.
- Slow reaction on restructuring and failures.

Example: Optimized Link State Routing Protocol (OLSR).

Distance Vector Routing

As in a fix net nodes maintain routing tables. Distance-vector protocols are based on calculating the direction and distance to any link in a network. "Direction" usually means the next hop address and the exit interface. "Distance" is a measure of the cost to reach a certain node. The least cost route between any two nodes is the route with minimum distance. Each node maintains a vector of minimum distance to every node. The cost of reaching a destination is calculated using various route metrics. RIP uses the hop count of the destination whereas IGRP takes into account other information such as node delay and available bandwidth.

Reactive Routing

This type of protocol finds a route based on user and traffic demand by flooding the network with Route Request or Discovery packets. The main disadvantages of such algorithms are:

- High latency time in route finding.
- Excessive flooding can lead to network clogging.

However, clustering can be used to limit flooding. The latency incurred during route discovery is not significant compared to periodic route update exchanges by all nodes in the network.

Example: Ad hoc On-Demand Distance Vector Routing (AODV).

Flooding

Is a simple routing algorithm in which every incoming packet is sent through every outgoing link except the one it arrived on? Flooding is used in bridging and in systems such as Usenet and peer-to-peer file sharing and as part of some routing protocols, including OSPF, DVMRP, and those used in wireless ad hoc networks.

Hybrid Routing

This type of protocol combines the advantages of proactive and reactive routing. The routing is initially established with some proactively prospected routes and then serves the demand from additionally activated nodes through reactive flooding. The choice of one or the other method requires predetermination for typical cases. The main disadvantages of such algorithms are:

- Advantage depends on number of other nodes activated.
- Reaction to traffic demand depends on gradient of traffic volume.

Example: Zone Routing Protocol (ZRP).

Position-based Routing

Position-based routing methods use information on the exact locations of the nodes. This information is obtained for example via a GPS receiver. Based on the exact location the best path between source and destination nodes can be determined.

Example: "Location-Aided Routing in mobile ad hoc networks" (LAR).

Technical Requirements for Implementation

An ad hoc network is made up of multiple "nodes" connected by "links."

Links are influenced by the node's resources (e.g., transmitter power, computing power and memory) and behavioral properties (e.g., reliability), as well as link properties (e.g. length-of-link and signal loss, interference and noise). Since links can be connected or disconnected at any time, a functioning network must be able to cope with this dynamic restructuring, preferably in a way that is timely, efficient, reliable, robust, and scalable.

The network must allow any two nodes to communicate by relaying the information via other nodes. A "path" is a series of links that connects two nodes. Various routing methods use one or two paths between any two nodes; flooding methods use all or most of the available paths.

Medium-access Control

In most wireless ad hoc networks, the nodes compete for access to shared wireless medium, often resulting in collisions (interference). Collisions can be handled using centralized scheduling or distributed contention access protocols. Using cooperative wireless communications improves immunity to interference by having the destination node combine self-interference and other-node interference to improve decoding of the desired signals.

Software Reprogramming

Large-scale ad hoc wireless networks may be deployed for long periods of time. During this time the requirements from the network or the environment in which the nodes are deployed may change. This can require modifying the application executing on the sensor nodes, or providing the application with a different set of parameters. It may be very difficult to manually reprogram the nodes because of the scale (possibly hundreds of nodes) and the embedded nature of the deployment, since the nodes may be located in places that are difficult to access physically. Therefore, the most relevant form of reprogramming is remote multihop reprogramming using the wireless medium which reprograms the nodes as they are embedded in their sensing environment. Specialized protocols have been developed for the embedded nodes which minimize the energy consumption of the process as well as reaching the entire network with high probability in as short a time as possible.

Simulation

One key problem in wireless ad hoc networks is foreseeing the variety of possible situations that can occur. As a result, modeling and simulation (M&S) using extensive parameter sweeping and what-if analysis becomes an extremely important paradigm for use in ad hoc networks. One solution is the use of simulation tools like OPNET, NetSim or ns2. A comparative study of various simulators for VANETs reveal that factors such as constrained road topology, multi-path fading and roadside obstacles, traffic flow models, trip models, varying vehicular speed and mobility, traffic lights, traffic congestion, drivers' behavior, etc., have to be taken into consideration in the simulation process to reflect realistic conditions.

Mathematical Models

The traditional model is the random geometric graph. Early work included simulating ad hoc mobile networks on sparse and densely connected topologies. Nodes are firstly scattered in a constrained physical space randomly. Each node then has a predefined fixed cell size (radio range). A node is said to be connected to another node if this neighbor is within its radio range. Nodes are then moved (migrated away) based on a random model, using random walk or brownian motion. Different mobility and number of nodes present yield different route length and hence different number of multi-hops.

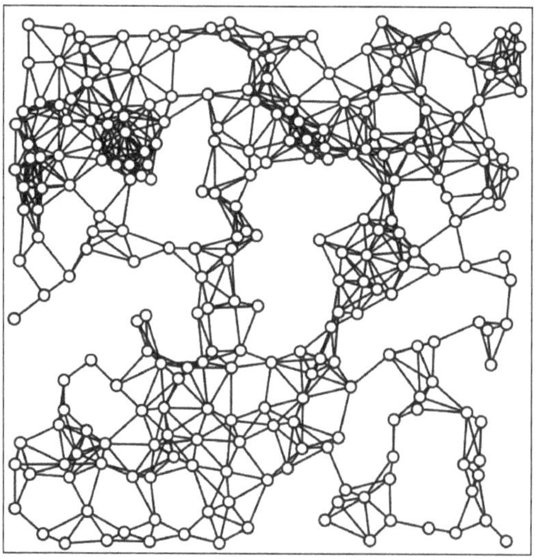

A randomly constructed geometric graph drawn inside a square.

These are graphs consisting of a set of nodes placed according to a point process in some usually bounded subset of the n-dimensional plane, mutually coupled according to a boolean probability mass function of their spatial separation. The connections between nodes may have different weights to model the difference in channel attenuations. One can then study network observables (such as connectivity, centrality or the degree distribution) from a graph-theoretic perspective. One can further study network protocols and algorithms to improve network throughput and fairness.

Security

Most wireless ad hoc networks do not implement any network access control, leaving these networks vulnerable to resource consumption attacks where a malicious node injects packets into the network with the goal of depleting the resources of the nodes relaying the packets.

To thwart or prevent such attacks, it was necessary to employ authentication mechanisms that ensure that only authorized nodes can inject traffic into the network. Even with authentication, these networks are vulnerable to packet dropping or delaying attacks, whereby an intermediate node drops the packet or delays it, rather than promptly sending it to the next hop.

Trust Management

Trust establishment and management in MANETs face challenges due to resource constraints and the complex interdependency of networks. Managing trust in a MANET needs to consider the interactions between the composite cognitive, social, information and communication networks, and take into account the resource constraints (e.g., computing power, energy, bandwidth, time), and dynamics (e.g., topology changes, node mobility, node failure, propagation channel conditions).

Researchers of trust management in MANET suggested that such complex interactions require a composite trust metric that captures aspects of communications and social networks, and corresponding trust measurement, trust distribution, and trust management schemes.

Wireless WAN

Wireless wide area network (WWAN), is a form of wireless network. The larger size of a wide area network compared to a local area network requires differences in technology. Wireless networks of different sizes deliver data in the form of telephone calls, web pages, and streaming video.

A WWAN often differs from wireless local area network (WLAN) by using mobile telecommunication cellular network technologies such as 2G, 3G, 4G LTE, and 5G to transfer data. It is sometimes referred as Mobile Broadband. These technologies are offered regionally, nationwide, or even globally and are provided by a wireless service provider. WWAN connectivity allows a user with a laptop and a WWAN card to surf the web, check email, or connect to a virtual private network (VPN) from anywhere within the regional boundaries of cellular service. Various computers can have integrated WWAN capabilities.

A WWAN may also be a closed network that covers a large geographic area. For example, a mesh network or MANET with nodes on buildings, towers, trucks, and planes could also be considered a WWAN.

A WWAN may also be a low-power, low-bit-rate wireless WAN, (LPWAN), intended to carry small packets of information between things, often in the form of battery operated sensors.

Since radio communications systems do not provide a physically secure connection path, WWANs typically incorporate encryption and authentication methods to make them more secure. Some of the early GSM encryption techniques were flawed, and security experts have issued warnings that cellular communication, including WWAN, is no longer secure. UMTS (3G) encryption was developed later and has yet to be broken.

Global Area Network

A global area network (GAN) refers to a network composed of different interconnected networks that cover an unlimited geographical area. The term is loosely synonymous with Internet, which is considered a global area network.

Unlike local area networks (LAN) and wide area networks (WAN), GANs cover a large geographical area.

The Global Area Network is the web connecting various terminals and LANs together. This is used so that the data can be transferred from one point to another even if they are not directly connected. For this type of network, there is either a central server or all connected terminals act as a relay for the data to find its way to the end point. Uses of many wireless connection and satellite coverage.

The mobile GAN is the use of terminals over a wide geographical zone to act as relays for wireless connection.

References

- Wireless-Networks: tutorialspoint.com, Retrieved 13 April, 2020
- Wireless-network: spiroprojects.com, Retrieved 09 June, 2020

- IEEE 802.11: Wireless LAN Medium Access Control (MAC) and Physical Layer (PHY) Specifications. (2016 revision). IEEE-SA. 14 December 2016. doi:10.1109/IEEESTD.2016.7786995. ISBN 978-1-5044-3645-8
- Dargie, W. and Poellabauer, C. (2010). Fundamentals of wireless sensor networks: theory and practice. John Wiley and Sons. pp. 168–183, 191–192. ISBN 978-0-470-99765-9
- Wireless-personal-area-network-wpan-5109: techopedia.com
- Wireless-PAN-Technologies, Wireless-PANs-Networks-for-Small-Places: etutorials.org, Retrieved 17 August, 2020
- Chai Keong Toh Ad Hoc Mobile Wireless Networks, Prentice Hall Publishers, 2002. ISBN 978-0-13-007817-9

3

Wireless Transmission and Technologies

Wireless transmission converts the digital data into wireless signals and spreads within its frequency range through the use of antennas. Radio transmission, infrared transmission, Bluetooth, Wi-Fi and microwave transmission are a few modes of transmission. This chapter has been carefully written to provide an easy understanding of wireless transmission and technologies.

Wireless Transmission

Wireless transmission is a form of unguided media. Wireless communication involves no physical link established between two or more devices, communicating wirelessly. Wireless signals are spread over in the air and are received and interpreted by appropriate antennas.

When an antenna is attached to electrical circuit of a computer or wireless device, it converts the digital data into wireless signals and spread all over within its frequency range. The receptor on the other end receives these signals and converts them back to digital data.

A little part of electromagnetic spectrum can be used for wireless transmission.

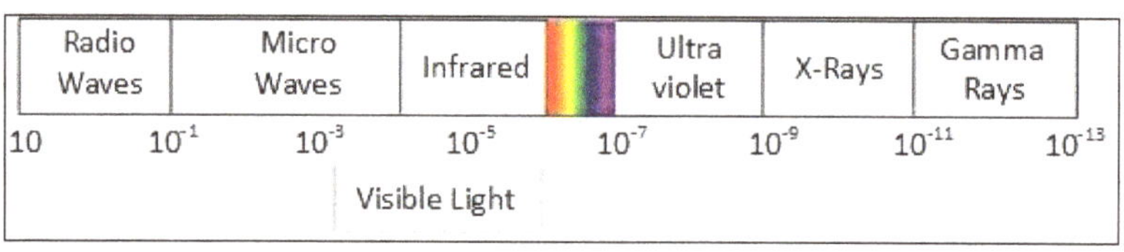

Radio Transmission

Radio frequency is easier to generate and because of its large wavelength it can penetrate through walls and structures alike. Radio waves can have wavelength from

1 mm – 100,000 km and have frequency ranging from 3 Hz (Extremely Low Frequency) to 300 GHz (Extremely High Frequency). Radio frequencies are sub-divided into six bands.

Radio waves at lower frequencies can travel through walls whereas higher RF can travel in straight line and bounce back. The power of low frequency waves decreases sharply as they cover long distance. High frequency radio waves have more power.

Lower frequencies such as VLF, LF, MF bands can travel on the ground up to 1000 kilometers, over the earth's surface.

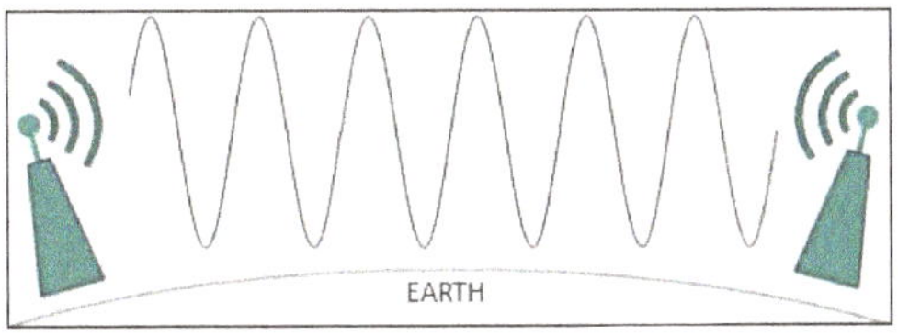

Radio waves of high frequencies are prone to be absorbed by rain and other obstacles. They use Ionosphere of earth atmosphere. High frequency radio waves such as HF and VHF bands are spread upwards. When they reach Ionosphere, they are refracted back to the earth.

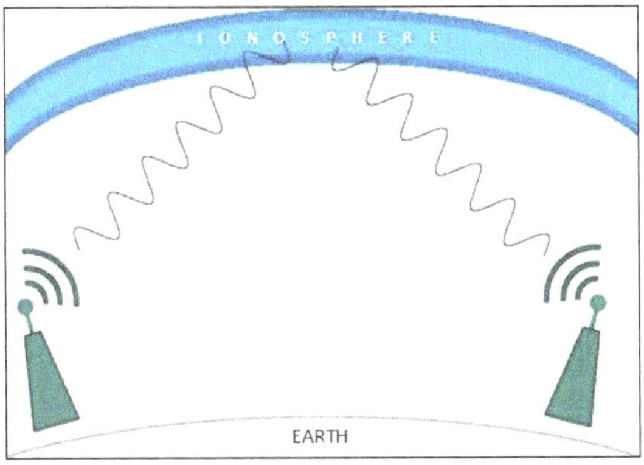

Microwave Transmission

Electromagnetic waves above 100 MHz tend to travel in a straight line and signals over them can be sent by beaming those waves towards one particular station. Because Microwaves travels in straight lines, both sender and receiver must be aligned to be strictly in line-of-sight.

Microwaves can have wavelength ranging from 1 mm – 1 meter and frequency ranging from 300 MHz to 300 GHz.

Microwave antennas concentrate the waves making a beam of it. Multiple antennas can be aligned to reach farther. Microwaves have higher frequencies and do not penetrate wall like obstacles.

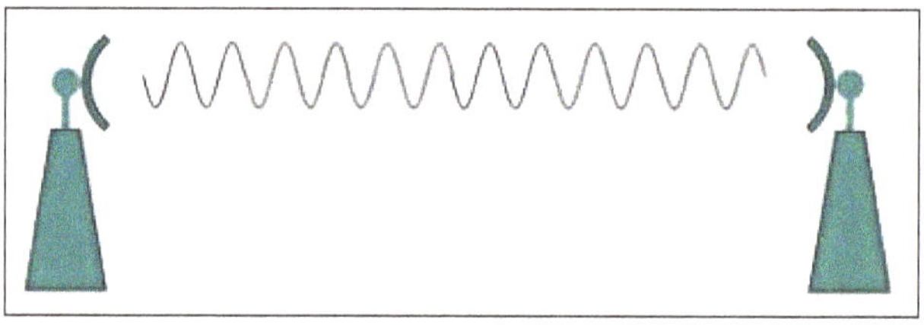

Microwave transmission depends highly upon the weather conditions and the frequency it is using.

Infrared Transmission

Infrared wave lies in between visible light spectrum and microwaves. It has wavelength of 700-nm to 1-mm and frequency ranges from 300-GHz to 430-THz.

Infrared wave is used for very short range communication purposes such as television and it's remote. Infrared travels in a straight line hence it is directional by nature. Because of high frequency range, Infrared cannot cross wall-like obstacles.

Light Transmission

Highest most electromagnetic spectrum which can be used for data transmission is light or optical signaling. This is achieved by means of LASER.

Because of frequency light uses, it tends to travel strictly in straight line. Hence the sender and receiver must be in the line-of-sight. Because laser transmission is unidirectional, at both ends of communication the laser and the photo-detector need to be installed. Laser beam is generally 1mm wide hence it is a work of precision to align two far receptors each pointing to lasers source.

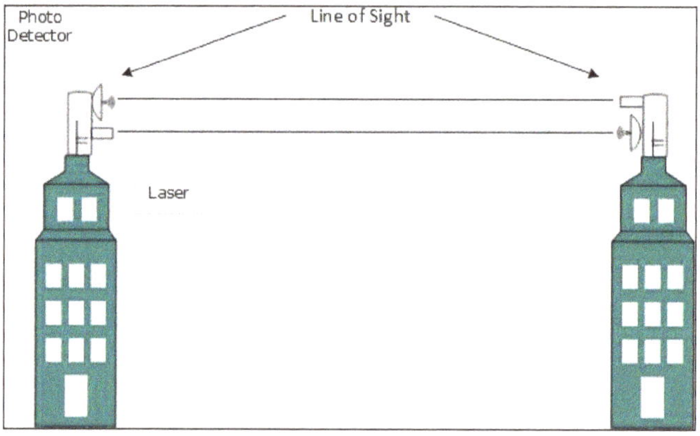

Laser works as Tx (transmitter) and photo-detectors works as Rx (receiver). Lasers cannot penetrate obstacles such as walls, rain, and thick fog. Additionally, laser beam is distorted by wind, atmosphere temperature, or variation in temperature in the path.

Laser is safe for data transmission as it is very difficult to tap 1mm wide laser without interrupting the communication channel.

Radio Transmitter

In electronics and telecommunications a transmitter or radio transmitter is an electronic device which produces radio waves with an antenna. The transmitter itself generates a radio frequency alternating current, which is applied to the antenna. When excited by this alternating current, the antenna radiates radio waves.

Transmitters are necessary component parts of all electronic devices that communicate by radio, such as radio and television broadcasting stations, cell phones, walkie-talkies, wireless computer networks, Bluetooth enabled devices, garage door openers, two-way radios in aircraft, ships, spacecraft, radar sets and navigational beacons. The term transmitter is usually limited to equipment that generates radio waves for communication purposes; or radiolocation, such as radar and navigational transmitters. Generators of radio waves for heating or industrial purposes, such as microwave ovens or diathermy equipment, are not usually called transmitters, even though they often have similar circuits.

The term is popularly used more specifically to refer to a broadcast transmitter, a transmitter used in broadcasting, as in FM radio transmitter or television transmitter. This usage typically includes the transmitter proper, the antenna, and often the building it is housed in.

A transmitter can be a separate piece of electronic equipment, or an electrical circuit within another electronic device. A transmitter and a receiver combined in one unit is called a transceiver. The term transmitter is often abbreviated "XMTR" or "TX" in technical documents. The purpose of most transmitters is radio communication of information over a distance. The information is provided to the transmitter in the form of an electronic signal, such as an audio (sound) signal from a microphone, a video (TV) signal from a video camera, or in wireless networking devices, and a digital signal from a computer. The transmitter combines the information signal to be carried with the radio frequency signal which generates the radio waves, which is called the carrier signal. This process is called modulation. The information can be added to the carrier in several different ways, in different types of transmitters. In an amplitude modulation (AM) transmitter, the information is added to the radio signal by varying its amplitude. In a frequency modulation (FM) transmitter, it is added by varying the radio signal's frequency slightly. Many other types of modulation are also used.

The radio signal from the transmitter is applied to the antenna, which radiates the energy as radio waves. The antenna may be enclosed inside the case or attached to the outside of the transmitter, as in portable devices such as cell phones, walkie-talkies, and garage door openers. In more powerful transmitters, the antenna may be located on top of a building or on a separate tower, and connected to the transmitter by a feed line, that is a transmission line.

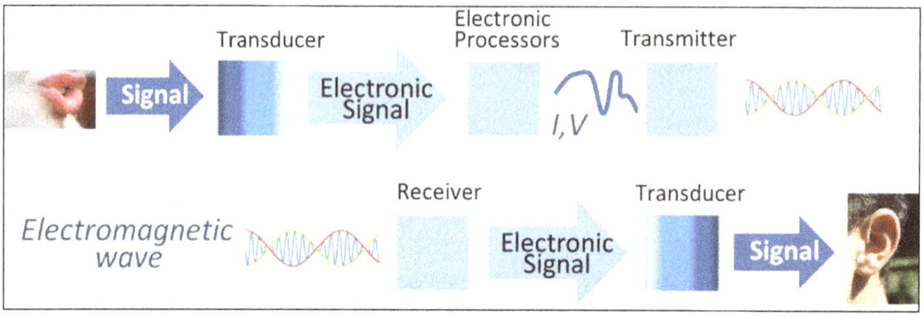

A radio transmitter is usually part of a radio communication system which uses electromagnetic waves (radio waves) to transport information (in this case sound) over a distance.

Operation

Electromagnetic waves are radiated by electric charges when they are accelerated. Radio waves, electromagnetic waves of radio frequency, are generated by time-varying electric currents, consisting of electrons flowing through a metal conductor called an antenna which are changing their velocity or direction and thus accelerating. An alternating current flowing back and forth in an antenna will create an oscillating magnetic field around the conductor. The alternating voltage will also charge the ends of the conductor alternately positive and negative, creating an oscillating electric field around the conductor. If the frequency of the oscillations is high enough, in the radio frequency range above about 20 kHz, the oscillating coupled electric and magnetic fields will radiate away from the antenna into space as an electromagnetic wave, a radio wave.

A radio transmitter is an electronic circuit which transforms electric power from a power source into a radio frequency alternating current to apply to the antenna, and the antenna radiates the energy from this current as radio waves. The transmitter also impresses information such as an audio or video signal onto the radio frequency current to be carried by the radio waves. When they strike the antenna of a radio receiver, the waves excite similar (but less powerful) radio frequency currents in it. The radio receiver extracts the information from the received waves.

Components

A practical radio transmitter mainly consists of the following parts:

- In high power transmitters, a power supply circuit to transform the input electrical power to the higher voltages needed to produce the required power output.
- An electronic oscillator circuit to generate the radio frequency signal. This usually generates a sine wave of constant amplitude called the carrier wave, because it serves to "carry" the information through space. In most modern transmitters, this is a crystal oscillator in which the frequency is precisely controlled by the vibrations of a quartz crystal. The frequency of the carrier wave is considered the frequency of the transmitter.

- A modulator circuit to add the information to be transmitted to the carrier wave produced by the oscillator. This is done by varying some aspect of the carrier wave. The information is provided to the transmitter either in the form of an audio signal, which represents sound, a video signal which represents moving images, or for data in the form of a binary digital signal which represents a sequence of bits, a bitstream. Different types of transmitters use different modulation methods to transmit information:
  - In an AM (amplitude modulation) transmitter the amplitude (strength) of the carrier wave is varied in proportion to the modulation signal.
  - In an FM (frequency modulation) transmitter the frequency of the carrier is varied by the modulation signal.
  - In an FSK (frequency-shift keying) transmitter, which transmits digital data, the frequency of the carrier is shifted between two frequencies which represent the two binary digits, 0 and 1.
  - OFDM (orthogonal frequency division multiplexing) is a family of complicated digital modulation methods very widely used in high bandwidth systems such as WiFi networks, cellphones, digital television broadcasting, and digital audio broadcasting (DAB) to transmit digital data using a minimum of radio spectrum bandwidth. OFDM has higher spectral efficiency and more resistance to fading than AM or FM. In OFDM multiple radio carrier waves closely spaced in frequency are transmitted within the radio channel, with each carrier modulated with bits from the incoming bitstream so multiple bits are being sent simultaneously, in parallel. At the receiver the carriers are demodulated and the bits are combined in the proper order into one bitstream.

  Many other types of modulation are also used. In large transmitters the oscillator and modulator together are often referred to as the exciter.

- A radio frequency (RF) amplifier to increase the power of the signal, to increase the range of the radio waves.
- An impedance matching (antenna tuner) circuit to match the impedance of the transmitter to the impedance of the antenna (or the transmission line to the antenna), to transfer power efficiently to the antenna. If these impedances are not equal, it causes a condition called standing waves, in which the power is reflected back from the antenna toward the transmitter, wasting power and sometimes overheating the transmitter.

In higher frequency transmitters, in the UHF and microwave range, free running oscillators are unstable at the output frequency. Older designs used an oscillator at a lower frequency, which was multiplied by frequency multipliers to get a signal at the desired frequency. Modern designs more commonly use an oscillator at the operating frequency which is stabilized by phase locking to a very stable lower frequency reference, usually a crystal oscillator.

Radio Receiver

In radio communications, a radio receiver, also known as a receiver, wireless or simply radio is an electronic device that receives radio waves and converts the information carried by them to a usable form. It is used with an antenna. The antenna intercepts radio waves (electromagnetic waves) and converts them to tiny alternating currents which are applied to the receiver, and the receiver extracts the desired information. The receiver uses electronic filters to separate the desired radio frequency signal from all the other signals picked up by the antenna, an electronic amplifier to increase the power of the signal for further processing, and finally recovers the desired information through demodulation.

Radio receivers are essential components of all systems that use radio. The information produced by the receiver may be in the form of sound, moving images (television), or digital data. A radio receiver may be a separate piece of electronic equipment, or an electronic circuit within another device. The most familiar type of radio receiver for most people is a broadcast radio receiver, which reproduces sound transmitted by radio broadcasting stations, historically the first mass-market radio application. A broadcast receiver is commonly called a "radio". However radio receivers are very widely used in other areas of modern technology, in televisions, cell phones, wireless modems and other components of communications, remote control, and wireless networking systems.

Broadcast Radio Receivers

The most familiar form of radio receiver is a broadcast receiver, often just called a radio, which receives audio programs intended for public reception transmitted by local radio stations. The sound is reproduced either by a loudspeaker in the radio or an earphone which plugs into a jack on the radio. The radio requires electric power, provided either by batteries inside the radio or a power cord which plugs into an electric outlet. All radios have a volume control to adjust the loudness of the audio, and some type of "tuning" control to select the radio station to be received.

Modulation Types

Modulation is the process of adding information to a radio carrier wave.

AM and FM

Two types of modulation are used in analog radio broadcasting systems; AM and FM.

In amplitude modulation (AM) the strength of the radio signal is varied by the audio signal. AM broadcasting is allowed in the AM broadcast bands which are between 148 and 283 kHz in the longwave range, and between 526 and 1706 kHz in the medium frequency (MF) range of the radio spectrum. AM broadcasting is also permitted in shortwave bands, between about 2.3 and 26 MHz, which are used for long distance international broadcasting.

In frequency modulation (FM) the frequency of the radio signal is varied slightly by the audio signal. FM broadcasting is permitted in the FM broadcast bands between about 65 and 108 MHz in the very high frequency (VHF) range. The exact frequency ranges vary somewhat in different countries.

FM stereo radio stations broadcast in stereophonic sound (stereo), transmitting two sound channels representing left and right microphones. A stereo receiver contains the additional circuits and parallel signal paths to reproduce the two separate channels. A monaural receiver, in contrast, only receives a single audio channel that is a combination (sum) of the left and right channels. While AM stereo transmitters and receivers exist, they have not achieved the popularity of FM stereo.

Most modern radios are "AM/FM" radios, and are able to receive both AM and FM radio stations, and have a switch to select which band to receive.

Digital Audio Broadcasting (DAB)

Digital audio broadcasting (DAB) is an advanced radio technology which debuted in some countries in 1998 that transmits audio from terrestrial radio stations as a digital signal rather than an analog signal as AM and FM do. Its advantages are that DAB has the potential to provide higher quality sound than FM (although many stations do not choose to transmit at such high quality), has greater immunity to radio noise and interference, makes better use of scarce radio spectrum bandwidth, and provides advanced user features such as electronic program guide, sports commentaries, and image slideshows. Its disadvantage is that it is incompatible with previous radios so that a new DAB receiver must be purchased. As of 2017, 38 countries offer DAB, with 2,100 stations serving listening areas containing 420 million people. Most countries plan an eventual switchover from FM to DAB. The United States and Canada have chosen not to implement DAB.

DAB radio stations work differently from AM or FM stations: a single DAB station transmits a wide 1,500 kHz bandwidth signal that carries from 9 to 12 channels from which the listener can choose. Broadcasters can transmit a channel at a range of different bit rates, so different channels can have different audio quality. In different countries DAB stations broadcast in either Band III (174–240 MHz) or L band (1.452–1.492 GHz).

Reception

The signal strength of radio waves decreases the farther they travel from the transmitter, so a radio station can only be received within a limited range of its transmitter. The range depends on the power of the transmitter, the sensitivity of the receiver, atmospheric and internal noise, as well as any geographical obstructions such as hills between transmitter and receiver. AM broadcast band radio waves travel as ground waves which follow the contour of the Earth, so AM radio stations can be reliably received at hundreds of miles distance. Due to their higher frequency, FM band radio signals cannot travel far beyond the visual horizon; limiting reception distance to about 40 miles (64 km), and can be blocked by hills between the transmitter and receiver. However FM radio is less susceptible to interference from radio noise (RFI, sferics, static) and has higher fidelity; better frequency response and less audio distortion, than AM. So in many countries serious music is only broadcast by FM stations, and AM stations specialize in radio news, talk radio, and sports. Like FM, DAB signals travel by line of sight so reception distances are limited by the visual horizon to about 30–40 miles (48–64 km).

Types of Broadcast Receiver

Radios are made in a range of styles and functions:

- Table radio: A self-contained radio with speaker designed to sit on a table.
- Clock radio: A bedside table radio that also includes an alarm clock. The alarm clock can be set to turn on the radio in the morning instead of an alarm, to wake the owner.
- Tuner: A high fidelity AM/FM radio receiver in a component home audio system. It has no speakers but outputs an audio signal which is fed into the system and played through the system's speakers.
- Portable radio: A radio powered by batteries that can be carried with a person. Radios are now often integrated with other audio sources in CD players and portable media players.
- Boom box: A portable battery-powered high fidelity stereo sound system in the form of a box with a handle, which became popular during the mid 1970s.
- Transistor radio: An older term for a portable pocket-sized broadcast radio receiver. Made possible by the invention of the transistor and developed in the 1950s, transistor radios were hugely popular during the 1960s and early 1970s, and changed the public's listening habits.
- Car radio: An AM/FM radio integrated into the dashboard of a vehicle, used for entertainment while driving. Virtually all modern cars and trucks are equipped with radios, which usually also includes a CD player.
- Satellite radio receiver: Subscription radio receiver that receives audio programming from a direct broadcast satellite. The subscriber must pay a monthly fee. They are mostly designed as car radios.
- Shortwave receiver: This is a broadcast radio that also receives the shortwave bands. It is used for shortwave listening.

- AV receivers are a common component in a high-fidelity or home-theatre system; in addition to receiving radio programming, the receiver will also contain switching and amplifying functions to interconnect and control the other components of the system.

A bedside clock radio that combines a radio receiver with an alarm clock.

Other Applications

Radio receivers are essential components of all systems that use radio. Besides broadcast receivers, radio receivers are used in a huge variety of electronic systems in modern technology. They can be a separate piece of equipment (a radio), or a subsystem incorporated into other electronic devices. A transceiver is a transmitter and receiver combined in one unit. Below is a list of a few of the most common types, organized by function:

- Broadcast television reception: Televisions receive a video signal representing a moving image, composed of a sequence of still images, and a synchronized audio signal representing the associated sound. The television channel received by a TV occupies a wider bandwidth than an audio signal, from 600 kHz to 6 MHz.
    - Terrestrial television receiver, broadcast television or just television (TV): Televisions contains an integral receiver (TV tuner) which receives free broadcast television from local television stations on TV channels in the VHF and UHF bands.
    - Satellite TV receiver: A set-top box which receives subscription direct-broadcast satellite television, and displays it on an ordinary television. A rooftop satellite dish receives many channels all modulated on a Ku band microwave downlink signal from a geostationary direct broadcast satellite 22,000 miles (35,000 km) above the Earth, and the signal is converted to a lower intermediate frequency and transported to the box through a coaxial cable. The subscriber pays a monthly fee.
- Two-way voice communications: A two-way radio is an audio transceiver, a receiver and transmitter in the same device, used for bidirectional person-to-person voice communication. The radio link may be half-duplex, using a single radio channel in which only one radio can transmit at a time. so

different users take turns talking, pressing a push to talk button on their radio which switches on the transmitter. Or the radio link may be full duplex, a bi-directional link using two radio channels so both people can talk at the same time, as in a cell phone.

- Cellphone: A portable telephone that is connected to the telephone network by radio signals exchanged with a local antenna called a cell tower. Cellphones have highly automated digital receivers working in the UHF and microwave band that receive the incoming side of the duplex voice channel, as well as a control channel that handles dialing calls and switching the phone between cell towers. They usually also have several other receivers that connect them with other networks: a WiFi modem, a bluetooth modem, and a GPS receiver. The cell tower has sophisticated multichannel receivers that receive the signals from many cell phones simultaneously.
- Cordless phone: A landline telephone in which the handset is portable and communicates with the rest of the phone by a short range duplex radio link, instead of being attached by a cord. Both the handset and the base station have radio receivers operating in the UHF band that receive the short range bidirectional duplex radio link.
- Citizens band radio: A two-way half-duplex radio operating in the 27 MHz band that can be used without a license. They are often installed in vehicles and used by truckers and delivery services.
- Walkie-talkie: A handheld short range half-duplex two-way radio.
- Scanner: A receiver that continuously monitors multiple frequencies or radio channels by stepping through the channels repeatedly, listening briefly to each channel for a transmission. When a transmitter is found the receiver stops at that channel. Scanners are used to monitor emergency police, fire, and ambulance frequencies, as well as other two way radio frequencies such as citizens band. Scanning capabilities have also become a standard feature in communications receivers, walkie-talkies, and other two-way radios.

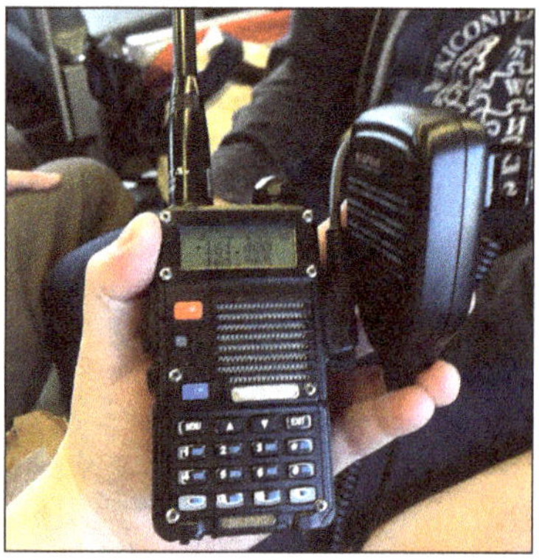

Handheld scanner.

- Communications receiver or shortwave receiver: A general purpose audio receiver covering the LF, MF, shortwave (HF), and VHF bands. Used mostly with a separate shortwave transmitter for two-way voice communication in communication stations, amateur radio stations, and for shortwave listening.

Modern communications receiver, ICOM RC-9500.

- One-way (simplex) voice communications:
  - Wireless microphone receiver: These receive the short range signal from wireless microphones used onstage by musical artists, public speakers, and television personalities.
  - Baby monitor: This is a crib side appliance for mothers of infants that transmits the baby's sounds to a receiver carried by the mother, so she can monitor the baby while she is in other parts of the house. Many baby monitors now have video cameras to show a picture of the baby.

Baby monitor. The receiver is on the left.

- Data communications:
  - Wireless (WiFi) modem: An automated short range digital data transmitter and receiver on a portable wireless device that communicates by microwaves with a nearby access point, a router or gateway, connecting the

portable device with a local computer network (WLAN) to exchange data with other devices.

    - Bluetooth modem: A very short range (up to 10 m) 2.4-2.83 GHz data transceiver on a portable wireless device used as a substitute for a wire or cable connection, mainly to exchange files between portable devices and connect cellphones and music players with wireless earphones.
    - Microwave relay: A long distance high bandwidth point-to-point data transmission link consisting of a dish antenna and transmitter that transmits a beam of microwaves to another dish antenna and receiver. Since the antennas must be in line-of-sight, distances are limited by the visual horizon to 30–40 miles. Microwave links are used for private business data, wide area computer networks (WANs), and by telephone companies to transmit distance phone calls and television signals between cities.
- Satellite communications: Communication satellites are used for data transmission between widely separated points on Earth. Other satellites are used for search and rescue, remote sensing, weather reporting and scientific research. Radio communication with satellites and spacecraft can involve very long path lengths, from 35,786 km (22,236 mi) for geosynchronous satellites to billions of kilometers for interplanetary spacecraft. This and the limited power available to a spacecraft transmitter mean very sensitive receivers must be used.
    - Satellite transponder: A receiver and transmitter in a communications satellite that receives multiple data channels carrying long distance telephone calls, television signals. or internet traffic on a microwave uplink signal from a satellite ground station and retransmits the data to another ground station on a different downlink frequency. In a direct broadcast satellite the transponder broadcasts a stronger signal directly to satellite radio or satellite television receivers in consumer's homes.
    - Satellite ground station receiver: Communication satellite ground stations receive data from communications satellites orbiting the Earth. Deep space ground stations such as those of the NASA Deep Space Network receive the weak signals from distant scientific spacecraft on interplanetary exploration missions. These have large dish antennas around 85 ft (25 m) in diameter, and extremely sensitive radio receivers similar to radio telescopes. The RF front end of the receiver is often cryogenically cooled to −195.79 °C (−320 °F) by liquid nitrogen to reduce radio noise in the circuit.
- Remote control: Remote control receivers receive digital commands that control a device, which may be as complex as a space vehicle or unmanned aerial vehicle, or as simple as a garage door opener. Remote control systems often also incorporate a telemetry channel to transmit data on the state of the controlled device back to the controller. Radio controlled model and other models include

multichannel receivers in model cars, boats, airplanes, and helicopters. A short-range radio system is used in keyless entry systems.

- Radiolocation: This is the use of radio waves to determine the location or direction of an object.
    - Radar: A device that transmits a narrow beam of microwaves which reflect from a target back to a receiver, used to locate objects such as aircraft, spacecraft, missiles, ships or land vehicles. The reflected waves from the target are received by a receiver usually connected to the same antenna, indicating the direction to the target. Widely used in aviation, shipping, navigation, weather forecasting, space flight, vehicle collision avoidance systems, and the military.
    - Global navigation satellite system (GNSS) receiver, such as a GPS receiver used with the US Global Positioning System - the most widely used electronic navigation device. An automated digital receiver that receives simultaneous data signals from several satellites in low Earth orbit. Using extremely precise time signals it calculates the distance to the satellites, and from this the receiver's location on Earth. GNSS receivers are sold as portable devices, and are also incorporated in cell phones, vehicles and weapons, even artillery shells.
    - VOR receiver: Navigational instrument on an aircraft that uses the VHF signal from VOR navigational beacons between 108 and 117.95 MHz to determine the direction to the beacon very accurately, for air navigation.
    - Wild animal tracking receiver: A receiver with a directional antenna used to track wild animals which have been tagged with a small VHF transmitter, for wildlife management purposes.
- Other:
    - Telemetry receiver: This receives data signals to monitor conditions of a process. Telemetry is used to monitor missile and spacecraft in flight, well logging during oil and gas drilling, and unmanned scientific instruments in remote locations.
    - Measuring receiver: A calibrated, laboratory grade radio receiver used to measure the characteristics of radio signals. Often incorporates a spectrum analyzer.
    - Radio telescope: Specialized antenna and radio receiver used as a scientific instrument to study weak radio waves from astronomical radio sources in space like stars, nebulas and galaxies in radio astronomy. They are the most sensitive radio receivers that exist, having large parabolic (dish) antennas up to 500 meters in diameter, and extremely sensitive radio circuits. The RF

front end of the receiver is often cryogenically cooled by liquid nitrogen to reduce radio noise.

Working Principle of Receivers

A radio receiver is connected to an antenna which converts some of the energy from the incoming radio wave into a tiny radio frequency AC voltage which is applied to the receiver's input. An antenna typically consists of an arrangement of metal conductors. The oscillating electric and magnetic fields of the radio wave push the electrons in the antenna back and forth, creating an oscillating voltage.

The antenna may be enclosed inside the receiver's case, as with the ferrite loop antennas of AM radios and the flat inverted F antenna of cell phones; attached to the outside of the receiver, as with whip antennas used on FM radios, or mounted separately and connected to the receiver by a cable, as with rooftop television antennas and satellite dishes.

Filtering, Amplification and Demodulation

Practical radio receivers perform three basic functions on the signal from the antenna: filtering, amplification, and demodulation:

- Bandpass filtering: Radio waves from many transmitters pass through the air simultaneously without interfering with each other. These can be separated in the receiver because they have different frequencies; that is, the radio wave from each transmitter oscillates at a different rate. To separate out the desired radio signal, the bandpass filter allows the frequency of the desired radio transmission to pass through, and blocks signals at all other frequencies.

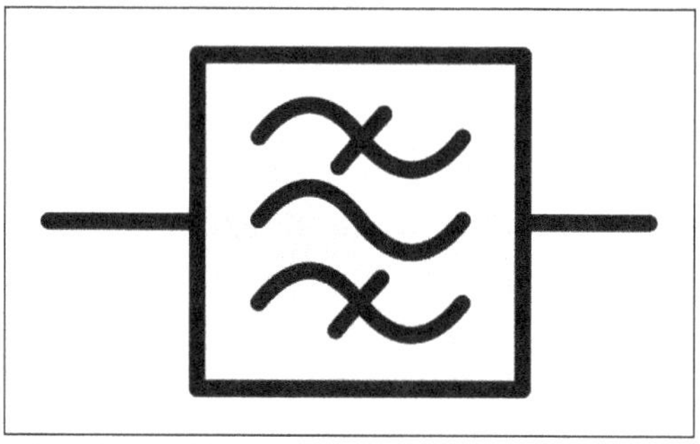

Symbol for a band pass filter used in block diagrams of radio receivers.

The bandpass filter consists of one or more resonant circuits (tuned circuits). The resonant circuit is connected between the antenna input and ground. When the incoming radio signal is at the resonant frequency, the resonant circuit has high impedance and the radio signal from the desired station is passed on to the following stages of the receiver. At all other frequencies the resonant circuit has low impedance, so signals at these frequencies are conducted to ground.

- Bandwidth and selectivity: The information (modulation) in a radio transmission is contained in two narrow bands of frequencies called sidebands (SB) on either side of the carrier frequency (C), so the filter has to pass a band of frequencies, not just a single frequency. The band of frequencies received by the receiver is called its passband (PB), and the width of the passband in kilohertz is called the bandwidth (BW). The bandwidth of the filter must be wide enough to allow the sidebands through without distortion, but narrow enough to block any interfering transmissions on adjacent frequencies. The ability of the receiver to reject unwanted radio stations near in frequency to the desired station is an important parameter called selectivity determined by the filter. In modern receivers quartz crystal, ceramic resonator, or surface acoustic wave (SAW) filters are often used which have sharper selectivity compared to networks of capacitor-inductor tuned circuits.
- Tuning: To select a particular station the radio is "tuned" to the frequency of the desired transmitter. The radio has a dial or digital display showing the frequency it is tuned to. Tuning is adjusting the frequency of the receiver's passband to the frequency of the desired radio transmitter. Turning the tuning knob changes the resonant frequency of the tuned circuit. When the resonant frequency is equal to the radio transmitter's frequency the tuned circuit oscillates in sympathy, passing the signal on to the rest of the receiver.

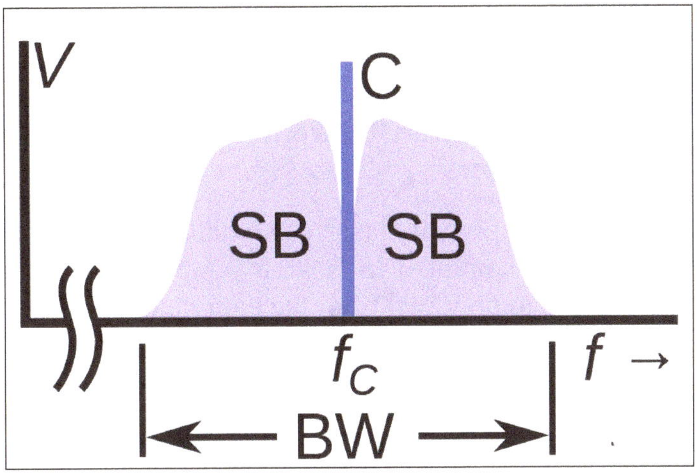

The frequency spectrum of a typical radio signal from an AM or FM radio transmitter. It consists of a component (C) at the carrier wave frequency f_C, with the modulation contained in narrow frequency bands called sidebands (SB) just above and below the carrier.

(right graph) How the bandpass filter selects a single radio signal S1 from all the radio signals received by the antenna. From top, the graphs show the voltage from the antenna applied to the filter Vin, the transfer function of the filter T, and the voltage at the output of the filter Vout as a function of frequency f. The transfer function T is the amount of signal that gets through the filter at each frequency.

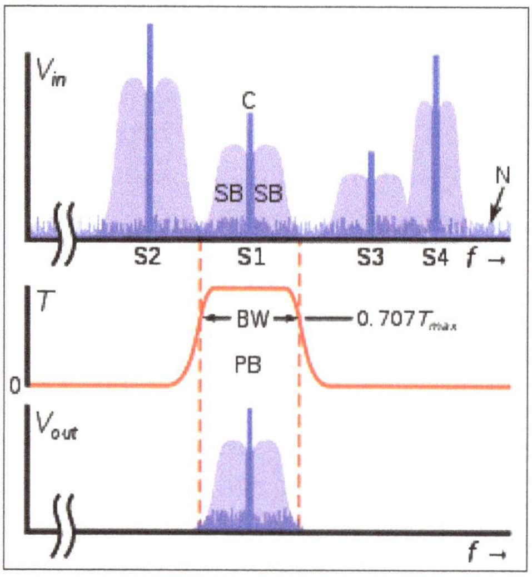

- Amplification: The power of the radio waves picked up by a receiving antenna decreases with the square of its distance from the transmitting antenna. Even with the powerful transmitters used in radio broadcasting stations, if the receiver is more than a few miles from the transmitter the power intercepted by the receiver's antenna is very small, perhaps as low as picowatts. To increase the power of the recovered signal, an amplifier circuit uses electric power from batteries or the wall plug to increase the amplitude (voltage or current) of the signal. In most modern receivers, the electronic components which do the actual amplifying are transistors.

  Receivers usually have several stages of amplification: The radio signal from the bandpass filter is amplified to make it powerful enough to drive the demodulator, then the audio signal from the demodulator is amplified to make it powerful enough to operate the speaker. The degree of amplification of a radio receiver is measured by a parameter called its sensitivity, which is the minimum signal strength of a station at the antenna, measured in microvolts, necessary to receive the signal clearly, with a certain signal-to-noise ratio. Since it is easy to amplify a signal to any desired degree, the limit to the sensitivity of many modern receivers is not the degree of amplification but random electronic noise present in the circuit, which can drown out a weak radio signal.

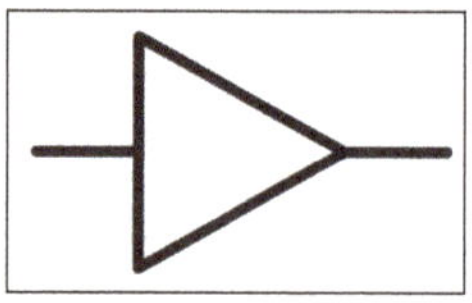

Symbol for an amplifier.

- Demodulation: After the radio signal is filtered and amplified, the receiver must extract the information-bearing modulation signal from the modulated radio frequency carrier wave. This is done by a circuit called a demodulator (detector). Each type of modulation requires a different type of demodulator:
    - An AM receiver that receives an (amplitude modulated) radio signal uses an AM demodulator.
    - An FM receiver that receives a frequency modulated signal uses an FM demodulator.
    - An FSK receiver which receives frequency shift keying (used to transmit digital data in wireless devices) uses an FSK demodulator.

Many other types of modulation are also used for specialized purposes.

The modulation signal output by the demodulator is usually amplified to increase its strength, and then the information is converted back to a human-usable form by some type of transducer. An audio signal, representing sound, as in a broadcast radio, is converted to sound waves by an earphone or loudspeaker. A video signal, representing moving images, as in a television receiver, is converted to light by a display. Digital data, as in a wireless modem, is applied as input to a computer or microprocessor, which interacts with human users.

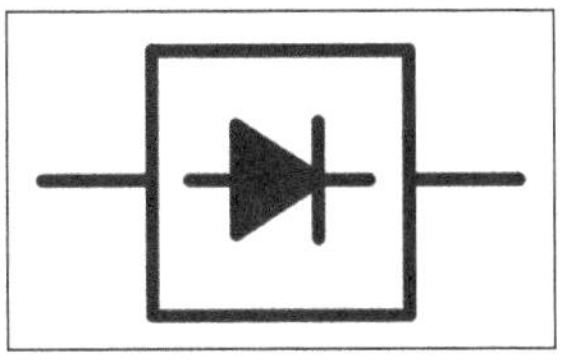

Symbol for a demodulator.

AM Demodulation

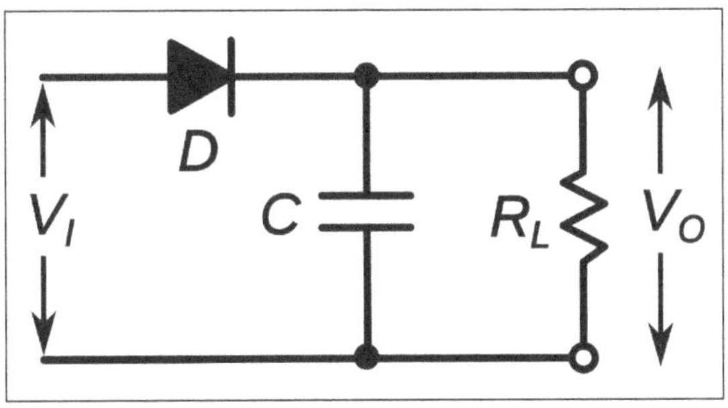

Envelope detector circuit.

The easiest type of demodulation to understand is AM demodulation, used in AM radios to recover the audio modulation signal, which represents sound and is converted to sound waves by the radio's speaker. It is accomplished by a circuit called an envelope detector, consisting of a diode (D) with a bypass capacitor (C) across its output.

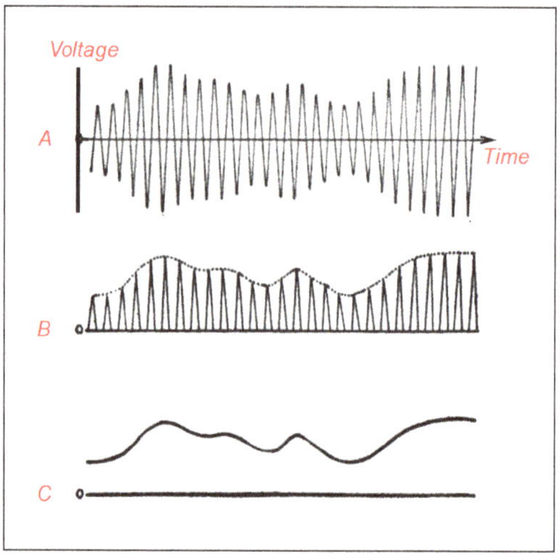

How an envelope detector works.

The amplitude modulated radio signal from the tuned circuit is shown at (A). The rapid oscillations are the radio frequency carrier wave. The audio signal (the sound) is contained in the slow variations (modulation) of the amplitude (size) of the waves. If it was applied directly to the speaker, this signal cannot be converted to sound, because the audio excursions are the same on both sides of the axis, averaging out to zero, which would result in no net motion of the speaker's diaphragm. (B) When this signal is applied as input VI to the detector, the diode (D) conducts current in one direction but not in the opposite direction, thus allowing through pulses of current on only one side of the signal. In other words, it rectifies the AC current to a pulsing DC current. The resulting voltage VO applied to the load RL no longer averages zero; its peak value is proportional to the audio signal. (C) The bypass capacitor (C) is charged up by the current pulses from the diode, and its voltage follows the peaks of the pulses, the envelope of the audio wave. It performs a smoothing (low pass filtering) function, removing the radio frequency carrier pulses, leaving the low frequency audio signal to pass through the load RL. The audio signal is amplified and applied to earphones or a speaker.

Tuned Radio Frequency (TRF) Receiver

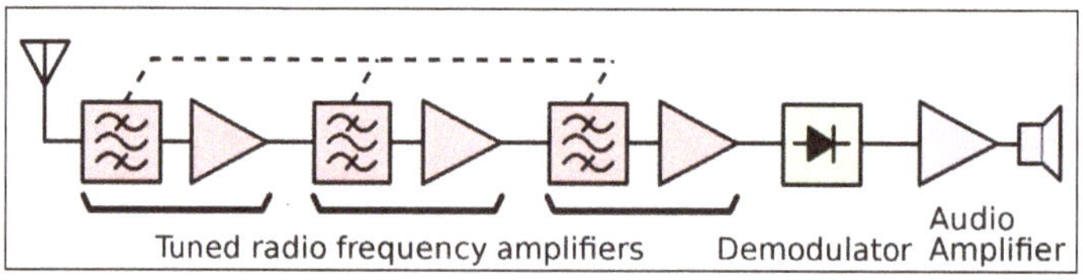

Block diagram of a tuned radio frequency receiver. To achieve enough selectivity to reject stations on adjacent frequencies, multiple cascaded bandpass filter stages had to be used. The dotted line indicates that the bandpass filters must be tuned together.

In the simplest type of radio receiver, called a tuned radio frequency (TRF) receiver, the three functions above are performed consecutively: (1) the mix of radio signals from the antenna is filtered to extract the signal of the desired transmitter; (2) this oscillating voltage is sent through a radio frequency (RF) amplifier to increase its strength to a level sufficient to drive the demodulator; (3) the demodulator recovers the modulation signal (which in broadcast receivers is an audio signal, a voltage oscillating at an audio frequency rate representing the sound waves) from the modulated radio carrier wave; (4) the modulation signal is amplified further in an audio amplifier, then is applied to a loudspeaker or earphone to convert it to sound waves.

Although the TRF receiver is used in a few applications, it has practical disadvantages which make it inferior to the superheterodyne receiver below, which is used in most applications. The drawbacks stem from the fact that in the TRF the filtering, amplification, and demodulation are done at the high frequency of the incoming radio signal. The bandwidth of a filter increases with its center frequency, so as the TRF receiver is tuned to different frequencies its bandwidth varies. Most important, the increasing congestion of the radio spectrum requires that radio channels be spaced very close together in frequency. It is extremely difficult to build filters operating at radio frequencies that have a narrow enough bandwidth to separate closely spaced radio stations. TRF receivers typically must have many cascaded tuning stages to achieve adequate selectivity.

Superheterodyne Design

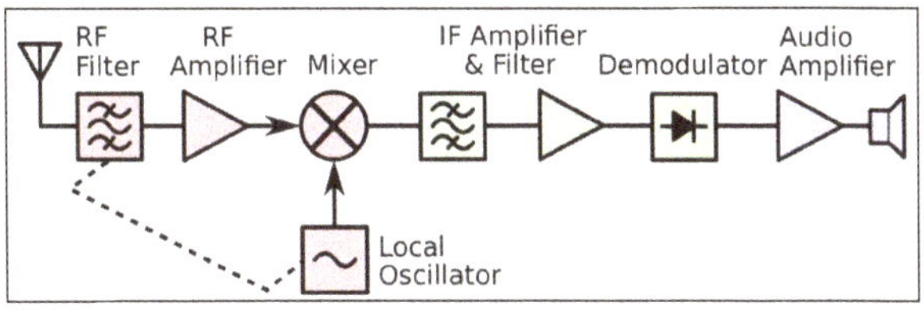

Block diagram of a superheterodyne receiver. The dotted line indicates that the RF filter and local oscillator must be tuned in tandem.

The superheterodyne receiver, invented in 1918 by Edwin Armstrong is the design used in almost all modern receivers except a few specialized applications.

In the superheterodyne, the radio frequency signal from the antenna is shifted down to a lower "intermediate frequency" (IF), before it is processed. The incoming radio frequency signal from the antenna is mixed with an unmodulated signal generated by a local oscillator (LO) in the receiver. The mixing is done in a nonlinear circuit called the "mixer". The result at the output of the mixer is a heterodyne or beat frequency at the difference between these two frequencies. The process is similar to the way two musical notes at different frequencies played together produce a beat note. This lower frequency is called the intermediate frequency (IF). The IF signal also has all the information that was present in the original RF signal. The IF signal passes through filter and amplifier stages, then is demodulated in a detector, recovering the original modulation.

The receiver is easy to tune; to receive a different frequency it is only necessary to change the local oscillator frequency. The stages of the receiver after the mixer operates at the fixed intermediate frequency (IF) so the IF bandpass filter does not have to be adjusted to different frequencies. The fixed frequency allows modern receivers to use sophisticated quartz crystal, ceramic resonator, or surface acoustic wave (SAW) IF filters that have very high Q factors, to improve selectivity.

The RF filter on the front end of the receiver is needed to prevent interference from any radio signals at the image frequency. Without an input filter the receiver can receive incoming RF signals at two different frequencies. The receiver can be designed to receive on either of these two frequencies; if the receiver is designed to receive on one, any other radio station or radio noise on the other frequency may pass through and interfere with the desired signal. A single tunable RF filter stage rejects the image frequency; since these are relatively far from the desired frequency, a simple filter provides adequate rejection. Rejection of interfering signals much closer in frequency to the desired signal is handled by the multiple sharply-tuned stages of the intermediate frequency amplifiers, which do no need to change their tuning. This filter does not need great selectivity, but as the receiver is tuned to different frequencies it must "track" in tandem with the local oscillator. The RF filter also serves to limit the bandwidth applied to the RF amplifier, preventing it from being overloaded by strong out-of-band signals.

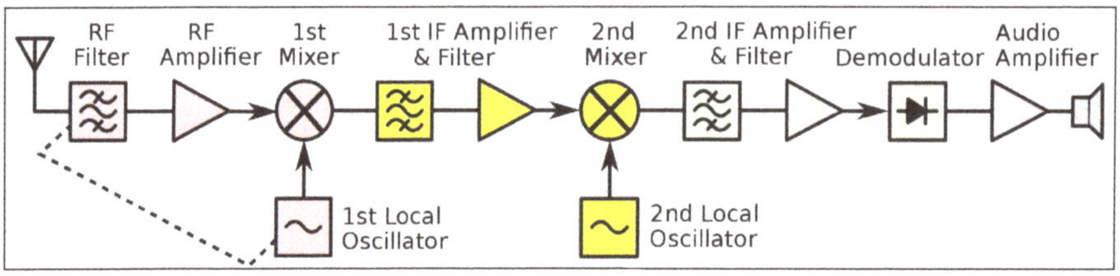

Block diagram of a dual-conversion superheterodyne receiver.

To achieve both good image rejection and selectivity, many modern superhet receivers use two intermediate frequencies; this is called a dual-conversion or double-conversion superheterodyne. The incoming RF signal is first mixed with one local oscillator signal in the first mixer to convert it to a high IF frequency, to allow efficient filtering out of the image frequency, then this first IF is mixed with a second local oscillator signal in a second mixer to convert it to a low IF frequency for good bandpass filtering. Some receivers even use triple-conversion.

At the cost of the extra stages, the superheterodyne receiver provides the advantage of greater selectivity than can be achieved with a TRF design. Where very high frequencies are in use, only the initial stage of the receiver needs to operate at the highest frequencies; the remaining stages can provide much of the receiver gain at lower frequencies which may be easier to manage. Tuning is simplified compared to a multi-stage TRF design, and only two stages need to track over the tuning range. The total amplification of the receiver is divided between three amplifiers at different frequencies; the RF, IF, and audio amplifier. This reduces problems with feedback and parasitic oscillations that are encountered in receivers where most of the amplifier stages operate at the same frequency, as in the TRF receiver.

The most important advantage is that better selectivity can be achieved by doing the filtering at the lower intermediate frequency. One of the most important parameters of a receiver is its bandwidth, the band of frequencies it accepts. In order to reject nearby interfering stations or noise, a narrow bandwidth is required. In all known filtering techniques, the bandwidth of the filter increases in proportion with the frequency, so by performing the filtering at the lower f_{IF}, rather than the frequency of the original radio signal f_{RF}, a narrower bandwidth can be achieved. Modern FM and television broadcasting, cellphones and other communications services, with their narrow channel widths, would be impossible without the superheterodyne.

Automatic Gain Control (AGC)

The signal strength (amplitude) of the radio signal from a receiver's antenna varies drastically, by orders of magnitude, depending on how far away the radio transmitter is, how powerful it is, and propagation conditions along the path of the radio waves. The strength of the signal received from a given transmitter varies with time due to changing propagation conditions of the path through which the radio wave passes, such as multipath interference; this is called fading. In an AM receiver the amplitude of the audio signal from the detector, and the sound volume, is proportional to the amplitude of the radio signal, so fading causes variations in the volume. In addition as the receiver is tuned between strong and weak stations, the volume of the sound from the speaker would vary drastically. Without an automatic system to handle it, in an AM receiver constant adjustment of the volume control would be required.

With other types of modulation like FM or FSK the amplitude of the modulation does not vary with the radio signal strength, but in all types the demodulator requires a certain range of signal amplitude to operate properly. Insufficient signal amplitude will cause an increase of noise in the demodulator, while excessive signal amplitude will cause amplifier stages to overload (saturate), causing distortion (clipping) of the signal.

Therefore, almost all modern receivers include a feedback control system which monitors the average level of the radio signal at the detector, and adjusts the gain of the amplifiers to give the optimum signal level for demodulation. This is called automatic gain control (AGC). AGC can be compared to the dark adaptation mechanism in the human eye; on entering a dark room the gain of the eye is increased by the iris opening. In its simplest form an AGC system consists of a rectifier which converts the RF signal to a varying DC level, a lowpass filter to smooth the variations and produce an average level. This is applied as a control signal to an earlier amplifier stage, to control its gain. In a superheterodyne receiver AGC is usually applied to the IF amplifier, and there may be a second AGC loop to control the gain of the RF amplifier to prevent it from overloading, too.

In certain receiver designs such as modern digital receivers, a related problem is DC offset of the signal. This is corrected by a similar feedback system.

Infrared Transmission

Infrared waves are those between the frequencies 300 GHz and 400 THz in the electromagnetic spectrum. Their wavelengths are shorter than microwaves but longer than visible light. Infrared propagation is line of sight.

They cannot penetrate walls and sun's infrared rays interfere with these rays. So cannot be used for long – range communication. As their usage is confined within closed space, they do not need any government permissions for their applications.

Applications of Infrared Waves in Communications

- Remote controls for television, stereos and other home appliances.
- Wireless LANs.
- Wireless modem, keyboard, mouse, printer etc.
- Fire detectors.
- Night vision systems.
- Intrusion detection systems.
- Motion detectors.

Microwave Transmission

Microwave transmission is the transmission of information by microwave radio waves. Although an experimental 40-mile (64 km) microwave telecommunication link across the English Channel was demonstrated in 1931, the development of radar in World War II provided the technology for practical exploitation of microwave communication. In the 1950s, large transcontinental microwave relay networks, consisting of chains of repeater stations linked by line-of-sight beams of microwaves were built in Europe and America to relay long distance telephone traffic and television programs between cities. Communication satellites which transferred data between ground stations by microwaves took over much long distance traffic in the 1960s. In recent years, there has been an explosive increase in use of the microwave spectrum by new telecommunication technologies such as wireless networks, and direct-broadcast satellites which broadcast television and radio directly into consumers' homes.

Microwaves are widely used for point-to-point communications because their small wavelength allows conveniently-sized antennas to direct them in narrow beams, which can be pointed directly at the receiving antenna. This allows nearby microwave equipment to use the same frequencies without interfering with each other, as lower frequency radio waves do. Another advantage is that the high frequency of microwaves gives the microwave band a very large information-carrying capacity; the microwave band has a bandwidth 30 times that of all the rest of the radio spectrum below it. A disadvantage is that microwaves are limited to line of sight propagation; they cannot pass around hills or mountains as lower frequency radio waves can.

A parabolic satellite antenna for Erdfunkstelle Raisting.

Microwave radio transmission is commonly used in point-to-point communication systems on the surface of the Earth, in satellite communications, and in deep space radio communications. Other parts of the microwave radio band are used for radars, radio navigation systems, sensor systems, and radio astronomy.

The next higher part of the radio electromagnetic spectrum, where the frequencies are above 30 GHz and below 100 GHz, are called "millimeter waves" because their wavelengths are conveniently measured in millimeters, and their wavelengths range from 10 mm down to 3.0 mm (Higher frequency waves are smaller in wavelength). Radio waves in this band are usually strongly attenuated by the Earthly atmosphere and particles contained in it, especially during wet weather. Also, in a wide band of frequencies around 60 GHz, the radio waves are strongly attenuated by molecular oxygen in the atmosphere. The electronic technologies needed in the millimeter wave band are also much more difficult to utilize than those of the microwave band.

Wireless Transmission of Information

- One-way (e.g. television broadcasting) and two-way telecommunication using communications satellite.
- Terrestrial microwave relay links in telecommunications networks including backbone or backhaul carriers in cellular networks.

Wireless Transmission of Power

- Proposed systems e.g. for connecting solar power collecting satellites to terrestrial power grids.

Microwave Radio Relay

C-band horn-reflector antennas on the roof of a telephone switching.

Dozens of microwave dishes on the Heinrich-Hertz-Turm.

Microwave radio relay is a technology widely used in the 1950s and 1960s for transmitting signals, such as long-distance telephone calls and television programs between two terrestrial points on a narrow beam of microwaves. In microwave radio relay, microwaves are transmitted on a line of sight path between relay stations using directional antennas, forming a fixed radio connection between the two points. The requirement of a line of sight limits the separation between stations to the visual horizon, about 30 to 50 miles (48 to 80 km). Before the widespread use of communications satellites, chains of microwave relay stations were used to transmit telecommunication signals over transcontinental distances.

Beginning in the 1950s, networks of microwave relay links, such as the AT&T Long Lines system in the U.S., carried long distance telephone calls and television programs between cities. The first system, dubbed TD-2 and built by AT&T, connected New York and Boston in 1947 with a series of eight radio relay stations. These included long daisy-chained series of such links that traversed mountain ranges and spanned continents. Much of the transcontinental traffic is now carried by cheaper optical fibers and communication satellites, but microwave relay remains important for shorter distances.

Planning

Communications tower with microwave relay dishes.

Because the radio waves travel in narrow beams confined to a line-of-sight path from one antenna to the other, they do not interfere with other microwave equipment, so nearby microwave links can use the same frequencies. Antennas must be highly directional (high gain); these antennas are installed in elevated locations such as large radio towers in order to be able to transmit across long distances. Typical types of antenna used in radio relay link installations are parabolic antennas, dielectric lens, and horn-reflector antennas, which have a diameter of up to 4 meters. Highly directive antennas permit an economical use of the available frequency spectrum, despite long transmission distances.

Danish military radio relay node.

Because of the high frequencies used, a line-of-sight path between the stations is required. Additionally, in order to avoid attenuation of the beam, an area around the beam called the first Fresnel zone must be free from obstacles. Obstacles in the signal field cause unwanted attenuation. High mountain peak or ridge positions are often ideal.

Production truck used for remote broadcasts by television news has a microwave dish on a retractible telescoping mast to transmit live video back to the studio.

Obstacles, the curvature of the Earth, the geography of the area and reception issues arising from the use of nearby land (such as in manufacturing and forestry) are important issues to consider when planning radio links. In the planning process, it is essential that "path profiles" are produced, which provide information about the terrain and Fresnel zones affecting the transmission path. The presence of a water surface, such as a lake or river, along the path also must be taken into consideration since it can reflect the beam, and the direct and reflected beam can interfere at the receiving antenna, causing multipath fading. Multipath fades are usually deep only in a small spot and a narrow frequency band, so space and/or frequency diversity schemes can be applied to mitigate these effects.

The effects of atmospheric stratification cause the radio path to bend downward in a typical situation so a major distance is possible as the earth equivalent curvature increases from 6370 km to about 8500 km (a 4/3 equivalent radius effect). Rare events of temperature, humidity and pressure profile versus height, may produce large deviations and distortion of the propagation and affect transmission quality. High-intensity rain and snow making rain fade must also be considered as an impairment factor, especially at frequencies above 10 GHz. All previous factors, collectively known as path loss, make it necessary to compute suitable power margins, in order to maintain the link operative for a high percentage of time, like the standard 99.99% or 99.999% used in 'carrier class' services of most telecommunication operators.

The longest microwave radio relay known up to date crosses the Red Sea with a 360 km (200 mi) hop between Jebel Erba (2170m a.s.l., 20°44′46.17″N 36°50′24.65″E, Sudan) and Jebel Dakka (2572m a.s.l., 21°5′36.89″N 40°17′29.80″E, Saudi Arabia). The link was built in 1979 by Telettra to transmit 300 telephone channels and one TV signal, in the 2 GHz frequency band. (Hop distance is the distance between two microwave stations).

Previous considerations represent typical problems characterizing terrestrial radio links using microwaves for the so-called backbone networks: hop lengths of a few tens of kilometers (typically 10 to 60 km) were largely used until the 1990s. Frequency bands below 10 GHz, and above all, the information to be transmitted, were a stream containing a fixed capacity block. The target was to supply the requested availability for the whole block (Plesiochronous digital hierarchy, PDH, or Synchronous Digital Hierarchy, SDH). Fading and/or multipath affecting the link for short time period during the day had to be counteracted by the diversity architecture. During 1990s microwave radio links begun widely to be used for urban links in cellular network. Requirements regarding link distance changed to shorter hops (less than 10 km, typically 3 to 5 km), and frequency increased to bands between 11 and 43 GHz and more recently, up to 86 GHz (E-band). Furthermore, link planning deals more with intense rainfall and less with multipath, so diversity schemes became less used. Another big change that occurred during the last decade was an evolution toward packet radio transmission. Therefore, new countermeasures, such as adaptive modulation, have been adopted.

The emitted power is regulated for cellular and microwave systems. These microwave transmissions use emitted power typically from 0.03 to 0.30 W, radiated by a parabolic antenna on a narrow beam diverging by a few degrees (1 to 3-4). The microwave channel arrangement is regulated by International Telecommunication Union (ITU-R) and local regulations (ETSI, FCC). In the last decade the dedicated spectrum for each microwave band has become extremely crowded, motivating the use of techniques to increase transmission capacity such as frequency reuse, Polarization-division multiplexing, XPIC, MIMO.

Microwave Link

A microwave link is a communications system that uses a beam of radio waves in the microwave frequency range to transmit video, audio, or data between two locations, which can be from just a few feet or meters to several miles or kilometers apart. Microwave links are commonly used by television broadcasters to transmit programmes across a country, for instance, or from an outside broadcast back to a studio.

Mobile units can be camera mounted, allowing cameras the freedom to move around without trailing cables. These are often seen on the touchlines of sports fields on Steadicam systems.

Properties of Microwave Links

- Involve line of sight (LOS) communication technology.
- Affected greatly by environmental constraints, including rain fade.
- Have very limited penetration capabilities through obstacles such as hills, buildings and trees.
- Sensitive to high pollen count.
- Signals can be degraded during solar proton events.

Uses of Microwave Links

- In communications between satellites and base stations.
- As backbone carriers for cellular systems.
- In short-range indoor communications.
- Linking remote and regional telephone exchanges to larger (main) exchanges without the need for copper/optical fiber lines.
- Measuring the intensity of rain between two locations.

Troposcatter

Terrestrial microwave relay links are limited in distance to the visual horizon, a few tens of miles or kilometers depending on tower height. Tropospheric scatter ("troposcatter" or "scatter") was a technology developed in the 1950s to allow microwave communication links beyond the horizon, to a range of several hundred kilometers. The transmitter radiates a beam of microwaves into the sky, at a shallow angle above the horizon toward the receiver. As the beam passes through the troposphere a small fraction of the microwave energy is scattered back toward the ground by water vapor and dust in the air. A sensitive receiver beyond the horizon picks up this reflected signal. Signal clarity obtained by this method depends on the weather and other factors, and as a result a high level of technical difficulty is involved in the creation of a reliable over horizon radio relay link. Troposcatter links are therefore only used in special circumstances where satellites and other long distance communication channels cannot be relied on, such as in military communications.

Bluetooth

Bluetooth is a wireless technology standard used for exchanging data between fixed and mobile devices over short distances using short-wavelength UHF radio waves in the industrial, scientific and medical radio bands, from 2.400 to 2.485 GHz, and building personal area networks (PANs). It was originally conceived as a wireless alternative to RS-232 data cables.

Bluetooth is managed by the Bluetooth Special Interest Group (SIG), which has more than 35,000 member companies in the areas of telecommunication, computing, networking, and consumer electronics. The IEEE standardized Bluetooth as IEEE 802.15.1, but no longer maintains the standard. The Bluetooth SIG oversees development of the specification, manages the qualification program, and protects the trademarks. A manufacturer must meet Bluetooth SIG standards to market it as a Bluetooth device. A network of patents apply to the technology, which are licensed to individual qualifying devices. As of 2009, Bluetooth integrated circuit chips ship approximately 920 million units annually.

Implementation

Bluetooth operates at frequencies between 2.402 and 2.480 GHz, or 2.400 and 2.4835 GHz including guard bands 2 MHz wide at the bottom end and 3.5 MHz wide at the top. This is in the globally unlicensed (but not unregulated) industrial, scientific and medical (ISM) 2.4 GHz short-range radio frequency band. Bluetooth uses a radio technology called frequency-hopping spread spectrum. Bluetooth divides transmitted data into packets, and transmits each packet on one of 79 designated Bluetooth channels. Each channel has a bandwidth of 1 MHz. It usually performs 1600 hops per second, with adaptive frequency-hopping (AFH) enabled. Bluetooth Low Energy uses 2 MHz spacing, which accommodates 40 channels.

Originally, Gaussian frequency-shift keying (GFSK) modulation was the only modulation scheme available. Since the introduction of Bluetooth 2.0+EDR, $\pi/4$-DQPSK (differential quadrature phase-shift keying) and 8-DPSK modulation may also be used between compatible devices. Devices functioning with GFSK are said to be operating in basic rate (BR) mode where an instantaneous bit rate of 1 Mbit/s is possible. The term Enhanced Data Rate (EDR) is used to describe π/4-DPSK and 8-DPSK schemes, each giving 2 and 3 Mbit/s respectively. The combination of these (BR and EDR) modes in Bluetooth radio technology is classified as a BR/EDR radio.

Bluetooth is a packet-based protocol with master/slave architecture. One master may communicate with up to seven slaves in a piconet. All devices share the master's clock. Packet exchange is based on the basic clock, defined by the master, which ticks at 312.5 μs intervals. Two clock ticks make up a slot of 625 μs, and two slots make up a slot pair of 1250 μs. In the simple case of single-slot packets, the master transmits in even slots and receives in odd slots. The slave, conversely, receives in even slots and transmits in odd slots. Packets may be 1, 3 or 5 slots long, but in all cases the master's transmission begins in even slots and the slave's in odd slots.

The above excludes Bluetooth Low Energy, introduced in the 4.0 specification, which uses the same spectrum but somewhat differently.

Communication and Connection

A master BR/EDR Bluetooth device can communicate with a maximum of seven devices in a piconet (an ad-hoc computer network using Bluetooth technology), though not all devices reach this maximum. The devices can switch roles, by agreement, and the slave can become the master (for example, a headset initiating a connection to a phone necessarily begins as master—as an initiator of the connection—but may subsequently operate as the slave).

The Bluetooth Core Specification provides for the connection of two or more piconets to form a scatternet, in which certain devices simultaneously play the master role in one piconet and the slave role in another.

At any given time, data can be transferred between the master and one other device (except for the little-used broadcast mode). The master chooses which slave device to address; typically, it switches rapidly from one device to another in a round-robin fashion. Since it is the master that chooses which slave to address, whereas a slave is (in theory) supposed to listen in each receive slot, being a master is a lighter burden than being a slave. Being a master of seven slaves is possible; being a slave of more than one master is possible. The specification is vague as to required behavior in scatternets.

Bluetooth is a standard wire-replacement communications protocol primarily designed for low power consumption, with a short range based on low-cost transceiver microchips in each device. Because the devices use a radio (broadcast) communications system, they do not have to be in visual line of sight of each other; however, a quasi-optical wireless path must be viable. Range is power-class-dependent, but effective ranges vary in practice.

Officially Class 3 radios have a range of up to 1 metre (3 ft.), Class 2, most commonly found in mobile devices, 10 metres (33 ft.), and Class 1, primarily for industrial use cases,100 metres (300 ft.). Bluetooth Marketing qualifies that Class 1 range is in most cases 20–30 metres (66–98 ft.) and Class 2 range 5–10 metres (16–33 ft.). The actual range achieved by a given link will depend on the qualities of the devices at both ends of the link, as well as the air conditions in between, and other factors.

The effective range varies depending on propagation conditions, material coverage, production sample variations, antenna configurations and battery conditions. Most Bluetooth applications are for indoor conditions, where attenuation of walls and signal fading due to signal reflections make the range far lower than specified line-of-sight ranges of the Bluetooth products.

Most Bluetooth applications are battery-powered Class 2 devices, with little difference in range whether the other end of the link is a Class 1 or Class 2 device as the lower-powered device tends to set the range limit. In some cases the effective range of the data link can be extended when a Class 2 device is connecting to a Class 1 transceiver with both higher sensitivity and transmission power than a typical Class 2 device. Mostly, however, the Class 1 devices have a similar sensitivity to Class 2 devices. Connecting two Class 1 devices with both high sensitivity and high power can allow ranges far in excess of the typical 100m, depending on the throughput required by the application. Some such devices allow open field ranges of up to 1 km and beyond between two similar devices without exceeding legal emission limits.

The Bluetooth Core Specification mandates a range of not less than 10 metres (33 ft.), but there is no upper limit on actual range. Manufacturers' implementations can be tuned to provide the range needed for each case.

Ranges of Bluetooth devices by class			
Class	Max. permitted power		Typ. Range (m)
	(mW)	(dBm)	
1	100	20	~100
1.5 (BT 5 Vol 6 Part A Sect 3)	10	10	~20
2	2.5	4	~10
3	1	0	~1
4	0.5	-3	~0.5

Bluetooth Profile

To use Bluetooth wireless technology, a device must be able to interpret certain Bluetooth profiles, which are definitions of possible applications and specify general behaviors that Bluetooth-enabled devices use to communicate with other Bluetooth devices. These profiles include settings to parameterize and to control the communication from the start. Adherence to profiles saves the time for transmitting the parameters anew before the bi-directional link becomes effective. There are a wide range of Bluetooth profiles that describe many different types of applications or use cases for devices.

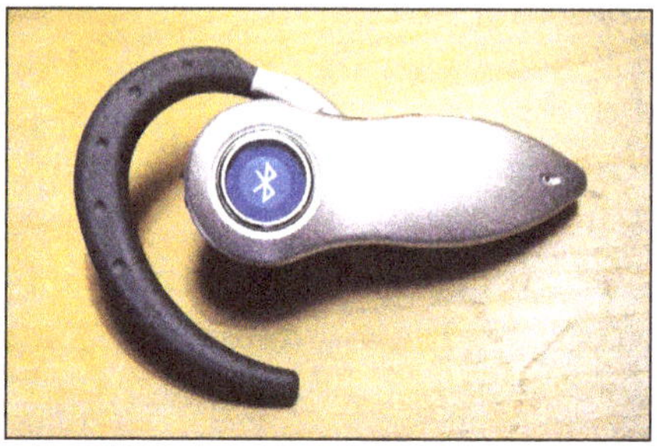

A typical Bluetooth mobile phone headset.

Bluetooth vs. Wi-Fi (IEEE 802.11)

Bluetooth and Wi-Fi (Wi-Fi is the brand name for products using IEEE 802.11 standards) have some similar applications: setting up networks, printing, or transferring files. Wi-Fi is intended as a replacement for high-speed cabling for general local area network access in work areas or home. This category of applications is sometimes called wireless local area networks (WLAN). Bluetooth was intended for portable equipment and its applications. The category of applications is outlined as the wireless personal area network (WPAN). Bluetooth is a replacement for cabling in a variety of personally carried applications in any setting, and also works for fixed location applications such as smart energy functionality in the home (thermostats, etc).

Wi-Fi and Bluetooth are to some extent complementary in their applications and usage. Wi-Fi is usually access point-centered, with an asymmetrical client-server connection with all traffic routed through the access point, while Bluetooth is usually symmetrical, between two Bluetooth devices. Bluetooth serves well in simple applications where two devices need to connect with a minimal configuration like a button press, as in headsets and remote controls, while Wi-Fi suits better in applications where some degree of client configuration is possible and high speeds are required, especially for network access through an access node. However, Bluetooth access points do exist, and ad-hoc connections are possible with Wi-Fi though not as simply as with Bluetooth. Wi-Fi Direct was recently developed to add more Bluetooth-like ad-hoc functionality to Wi-Fi.

Devices

Bluetooth exists in numerous products such as telephones, speakers, tablets, media players, robotics systems, laptops, and console gaming equipment as well as some high definition headsets, modems, hearing aids and even watches. Given the variety of devices which use the Bluetooth, coupled with the contemporary deprecation of headphone jacks by Apple, Google, and other companies, and the lack of regulation by the FCC, the technology is prone to interference. Nonetheless Bluetooth is useful when transferring information between two or more devices that are near each other in low-bandwidth situations. Bluetooth is commonly used to transfer sound data with telephones (i.e., with a Bluetooth headset) or byte data with hand-held computers (transferring files).

Bluetooth protocols simplify the discovery and setup of services between devices. Bluetooth devices can advertise all of the services they provide. This makes using services easier, because more of the security, network address and permission configuration can be automated than with many other network types.

A Bluetooth USB dongle with a 100 m range.

Computer Requirements

A personal computer that does not have embedded Bluetooth can use a Bluetooth adapter that enables the PC to communicate with Bluetooth devices. While some desktop computers and most recent laptops come with a built-in Bluetooth radio, others require an external adapter, typically in the form of a small USB "dongle."

A typical Bluetooth USB dongle.

An internal notebook Bluetooth card (14×36×4 mm).

Unlike its predecessor, IrDA, which requires a separate adapter for each device, Bluetooth lets multiple devices communicate with a computer over a single adapter.

Operating System Implementation

For Microsoft platforms, Windows XP Service Pack 2 and SP3 releases work natively with Bluetooth v1.1, v2.0 and v2.0+EDR. Previous versions required users to install their Bluetooth adapter's own drivers, which were not directly supported by Microsoft. Microsoft's own Bluetooth dongles (packaged with their Bluetooth computer devices) have no external drivers and thus require at least Windows XP Service Pack 2. Windows Vista RTM/SP1 with the Feature Pack for Wireless or Windows Vista SP2 work with Bluetooth v2.1+EDR. Windows 7 works with Bluetooth v2.1+EDR and Extended Inquiry Response (EIR). The Windows XP and Windows Vista/Windows 7 Bluetooth stacks support the following Bluetooth profiles natively: PAN, SPP, DUN, HID, HCRP. The Windows XP stack can be replaced by a third party stack that supports more profiles or newer Bluetooth versions. The Windows Vista/Windows 7 Bluetooth stack supports vendor-supplied additional profiles without requiring that the Microsoft stack be replaced. It is generally recommended to install the latest vendor driver and its associated stack to be able to use the Bluetooth device at its fullest extent.

Apple products have worked with Bluetooth since Mac OS X v10.2, which was released in 2002.

Linux has two popular Bluetooth stacks, BlueZ and Fluoride. The BlueZ stack is included with most Linux kernels and was originally developed by Qualcomm. Fluoride, earlier known as Bluedroid is included in Android OS and was originally developed by Broadcom. There is also Affix stack, developed by Nokia. It was once popular, but has not been updated since 2005.

FreeBSD has included Bluetooth since its v5.0 release, implemented through netgraph.

NetBSD has included Bluetooth since its v4.0 release. Its Bluetooth stack was ported to OpenBSD as well, however OpenBSD later removed it as unmaintained.

DragonFly BSD has had NetBSD's Bluetooth implementation since 1.11. A netgraph-based implementation from FreeBSD has also been available in the tree, possibly disabled until 2014-11-15, and may require more work.

Specifications and Features

The specifications were formalized by the Bluetooth Special Interest Group (SIG) and formally announced on the 20 of May 1998. Today it has a membership of over 30,000 companies worldwide. It was established by Ericsson, IBM, Intel, Nokia and Toshiba, and later joined by many other companies.

All versions of the Bluetooth standards support downward compatibility. That lets the latest standard cover all older versions.

The Bluetooth Core Specification Working Group (CSWG) produces mainly 4 kinds of specifications:

- The Bluetooth Core Specification, release cycle is typically a few years in between.
- Core Specification Addendum (CSA), release cycle can be as tight as a few times per year.
- Core Specification Supplements (CSS), can be released very quickly.
- Errata Available with a user account: Errata login.

Technical Information and Architecture

Software

Seeking to extend the compatibility of Bluetooth devices, the devices that adhere to the standard use an interface called HCI (Host Controller Interface) between the host device (laptop, phone, etc.) and the Bluetooth device as such (Bluetooth chip).

High-level protocols such as the SDP (Protocol used to find other Bluetooth devices within the communication range, also responsible for detecting the function of devices in range), RFCOMM (Protocol used to emulate serial port connections) and TCS (Telephony control protocol) interact with the baseband controller through the L2CAP Protocol (Logical Link Control and Adaptation Protocol). The L2CAP protocol is responsible for the segmentation and reassembly of the packets.

Hardware

The hardware that makes up the Bluetooth device is made up of, logically, two parts; which may or may not be physically separate. A radio device, responsible for modulating and transmitting the signal; and a digital controller. The digital controller is likely a CPU, one of whose functions is to run a Link Controller; and interfaces with the host device; but some functions may be delegated to hardware. The Link Controller is responsible for the processing of the baseband and the management of ARQ and physical layer FEC protocols. In addition, it handles the transfer functions (both asynchronous and synchronous), audio coding and data encryption. The CPU of the device is responsible for attending the instructions related to Bluetooth of the host device, in order to simplify its operation. To do this, the CPU runs software called Link Manager that has the function of communicating with other devices through the LMP protocol.

A Bluetooth device is a short-range wireless device. Bluetooth devices are fabricated on RF CMOS integrated circuit (RF circuit) chips.

Bluetooth Protocol Stack

Bluetooth is defined as a layer protocol architecture consisting of core protocols, cable replacement protocols, telephony control protocols, and adopted protocols. Mandatory protocols for all Bluetooth stacks are LMP, L2CAP and SDP. In addition, devices that communicate with Bluetooth almost universally can use these protocols: HCI and RFCOMM.

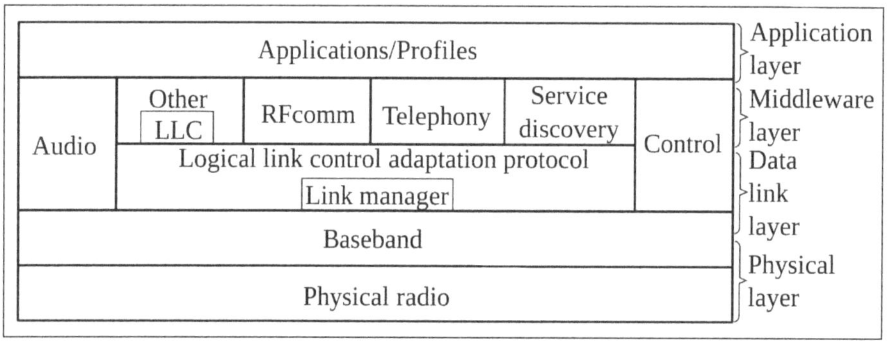

Bluetooth Protocol Stack.

Link Manager

The Link Manager (LM) is the system that manages establishing the connection between devices. It is responsible for the establishment, authentication and configuration of the link. The Link Manager locates other managers and communicates with them via the management protocol of the LMP link. In order to perform its function as a service provider, the LM uses the services included in the Link Controller (LC). The Link Manager Protocol basically consists of a number of PDUs (Protocol Data Units) that are sent from one device to another. The following is a list of supported services:

- Transmission and reception of data.
- Name request.
- Request of the link addresses.
- Establishment of the connection.
- Authentication.
- Negotiation of link mode and connection establishment.

Host Controller Interface

The Host Controller Interface provides a command interface for the controller and for the link manager, which allows access to the hardware status and control registers. This interface provides an access layer for all Bluetooth devices. The HCI layer of the machine exchanges commands and data with the HCI firmware present in the Bluetooth device. One of the most important HCI tasks that must be performed is the automatic discovery of other Bluetooth devices that are within the coverage radius.

Logical Link Control and Adaptation Protocol

The Logical Link Control and Adaptation Protocol (L2CAP) is used to multiplex multiple logical connections between two devices using different higher level protocols. Provides segmentation and reassembly of on-air packets.

In Basic mode, L2CAP provides packets with a payload configurable up to 64 kB, with 672 bytes as the default MTU, and 48 bytes as the minimum mandatory supported MTU.

In Retransmission and Flow Control modes, L2CAP can be configured either for isochronous data or reliable data per channel by performing retransmissions and CRC checks.

Bluetooth Core Specification Addendum 1 adds two additional L2CAP modes to the core specification. These modes effectively deprecate original Retransmission and Flow Control modes.

Enhanced Retransmission Mode (ERTM)

This mode is an improved version of the original retransmission mode. This mode provides a reliable L2CAP channel.

Streaming Mode (SM)

This is a very simple mode, with no retransmission or flow control. This mode provides an unreliable L2CAP channel.

Reliability in any of these modes is optionally and/or additionally guaranteed by the lower layer Bluetooth BDR/EDR air interface by configuring the number of retransmissions and flush timeout (time after which the radio flushes packets). In-order sequencing is guaranteed by the lower layer.

Only L2CAP channels configured in ERTM or SM may be operated over AMP logical links.

Service Discovery Protocol

The Service Discovery Protocol (SDP) allows a device to discover services offered by other devices, and their associated parameters. For example, when you use a mobile phone with a Bluetooth headset, the phone uses SDP to determine which Bluetooth profiles the headset can use (Headset Profile, Hands Free Profile, Advanced Audio Distribution Profile (A2DP) etc.) and the protocol multiplexer settings needed for the phone to connect to the headset using each of them. Each service is identified by a Universally Unique Identifier (UUID), with official services (Bluetooth profiles) assigned a short form UUID (16 bits rather than the full 128).

Radio Frequency Communications

Radio Frequency Communications (RFCOMM) is a cable replacement protocol used for generating a virtual serial data stream. RFCOMM provides for binary data transport and emulates EIA-232 (formerly RS-232) control signals over the Bluetooth baseband layer, i.e., it is a serial port emulation.

RFCOMM provides a simple, reliable, data stream to the user, similar to TCP. It is used directly by many telephony related profiles as a carrier for AT commands, as well as being a transport layer for OBEX over Bluetooth.

Many Bluetooth applications use RFCOMM because of its widespread support and publicly available API on most operating systems. Additionally, applications that used a serial port to communicate can be quickly ported to use RFCOMM.

Bluetooth Network Encapsulation Protocol

The Bluetooth Network Encapsulation Protocol (BNEP) is used for transferring another protocol stack's data via an L2CAP channel. Its main purpose is the transmission of IP packets in the Personal Area Networking Profile. BNEP performs a similar function to SNAP in Wireless LAN.

Audio/Video Control Transport Protocol

The Audio/Video Control Transport Protocol (AVCTP) is used by the remote control profile to transfer AV/C commands over an L2CAP channel. The music control buttons on a stereo headset use this protocol to control the music player.

Audio/Video Distribution Transport Protocol

The Audio/Video Distribution Transport Protocol (AVDTP) is used by the advanced audio distribution (A2DP) profile to stream music to stereo headsets over an L2CAP channel intended for video distribution profile in the Bluetooth transmission.

Telephony Control Protocol

The Telephony Control Protocol – Binary (TCS BIN) is the bit-oriented protocol that defines the call control signaling for the establishment of voice and data calls between Bluetooth devices. Additionally, "TCS BIN defines mobility management procedures for handling groups of Bluetooth TCS devices."

TCS-BIN is only used by the cordless telephony profile, which failed to attract implementers. As such it is only of historical interest.

Adopted Protocols

Adopted protocols are defined by other standards-making organizations and incorporated into Bluetooth's protocol stack, allowing Bluetooth to code protocols only when necessary. The adopted protocols include:

- Point-to-Point Protocol (PPP): Internet standard protocol for transporting IP datagrams over a point-to-point link.
- TCP/IP/UDP: Foundation Protocols for TCP/IP protocol suite.
- Object Exchange Protocol (OBEX): Session-layer protocol for the exchange of objects, providing a model for object and operation representation.
- Wireless Application Environment/Wireless Application Protocol (WAE/WAP): WAE specifies an application framework for wireless devices and WAP is an open standard to provide mobile users access to telephony and information services.
- Baseband Error Correction: Depending on packet type, individual packets may be protected by error correction, either 1/3 rate forward error correction (FEC) or 2/3 rate. In addition, packets with CRC will be retransmitted until acknowledged by automatic repeat request (ARQ).

Setting up Connections

Any Bluetooth device in discoverable mode transmits the following information on demand:

- Device name.
- Device class.
- List of services.
- Technical information (for example: device features, manufacturer, Bluetooth specification used, clock offset).

Any device may perform an inquiry to find other devices to connect to, and any device can be configured to respond to such inquiries. However, if the device trying to connect knows the address of the device, it always responds to direct connection requests and transmits the information shown in the list above if requested. Use of a device's services may require pairing or acceptance by its owner, but the connection itself can be initiated by any device and held until it goes out of range. Some devices can be connected to only one device at a time, and connecting to them prevents them from connecting to other devices and appearing in inquiries until they disconnect from the other device.

Every device has a unique 48-bit address. However, these addresses are generally not shown in inquiries. Instead, friendly Bluetooth names are used, which can be set by the user. This name appears when another user scans for devices and in lists of paired devices.

Most cellular phones have the Bluetooth name set to the manufacturer and model of the phone by default. Most cellular phones and laptops show only the Bluetooth names and special programs are required to get additional information about remote devices. This can be confusing as, for example, there could be several cellular phones in range named T610.

Pairing and Bonding

Motivation

Many services offered over Bluetooth can expose private data or let a connecting party control the Bluetooth device. Security reasons make it necessary to recognize specific devices, and thus enable control over which devices can connect to a given Bluetooth device. At the same time, it is useful for Bluetooth devices to be able to establish a connection without user intervention (for example, as soon as in range).

To resolve this conflict, Bluetooth uses a process called bonding, and a bond is generated through a process called pairing. The pairing process is triggered either by a specific request from a user to generate a bond (for example, the user explicitly requests to "Add a Bluetooth device"), or it is triggered automatically when connecting to a service where (for the first time) the identity of a device is required for security purposes. These two cases are referred to as dedicated bonding and general bonding respectively.

Pairing often involves some level of user interaction. This user interaction confirms the identity of the devices. When pairing successfully completes, a bond forms between the two devices, enabling those two devices to connect to each other in the future without repeating the pairing process to confirm device identities. When desired, the user can remove the bonding relationship.

Implementation

During pairing, the two devices establish a relationship by creating a shared secret known as a link key. If both devices store the same link key, they are said to be paired or bonded. A device that wants to communicate only with a bonded device can cryptographically authenticate the identity of the other device, ensuring it is the same device it previously paired with. Once a link key is generated, an authenticated Asynchronous Connection-Less (ACL) link between the devices may be encrypted to protect exchanged data against eavesdropping. Users can delete link keys from either device, which removes the bond between the devices—so it is possible for one device to have a stored link key for a device it is no longer paired with.

Bluetooth services generally require either encryption or authentication and as such require pairing before they let a remote device connect. Some services, such as the Object Push Profile, elect not to explicitly require authentication or encryption so that pairing does not interfere with the user experience associated with the service use-cases.

Pairing Mechanisms

Pairing mechanisms changed significantly with the introduction of Secure Simple Pairing in Bluetooth v2.1. The following summarizes the pairing mechanisms:

- Legacy pairing: This is the only method available in Bluetooth v2.0 and before. Each device must enter a PIN code; pairing is only successful if both devices enter the same PIN code. Any 16-byte UTF-8 string may be used as a PIN code; however, not all devices may be capable of entering all possible PIN codes.
    - Limited input devices: The obvious example of this class of device is a Bluetooth Hands-free headset, which generally have few inputs. These devices usually have a fixed PIN, for example "0000" or "1234", that are hard-coded into the device.
    - Numeric input devices: Mobile phones are classic examples of these devices. They allow a user to enter a numeric value up to 16 digits in length.
    - Alpha-numeric input devices: PCs and smartphones are examples of these devices. They allow a user to enter full UTF-8 text as a PIN code. If pairing with a less capable device the user must be aware of the input limitations on the other device; there is no mechanism available for a capable device to determine how it should limit the available input a user may use.
- Secure Simple Pairing (SSP): This is required by Bluetooth v2.1, although a Bluetooth v2.1 device may only use legacy pairing to interoperate with a v2.0 or earlier device. Secure Simple Pairing uses a form of public key cryptography, and some types can help protect against man in the middle, or MITM attacks. SSP has the following authentication mechanisms:
    - Just works: As the name implies, this method just works, with no user interaction. However, a device may prompt the user to confirm the pairing process. This method is typically used by headsets with very limited IO capabilities, and is more secure than the fixed PIN mechanism this limited set of devices uses for legacy pairing. This method provides no man-in-the-middle (MITM) protection.
    - Numeric comparison: If both devices have a display, and at least one can accept a binary yes/no user input, they may use Numeric Comparison. This method displays a 6-digit numeric code on each device. The user should compare the numbers to ensure they are identical. If the comparison succeeds, the user(s) should confirm pairing on the device(s) that can accept an input. This method provides MITM protection, assuming the user confirms on both devices and actually performs the comparison properly.
    - Passkey Entry: This method may be used between a device with a display and a device with numeric keypad entry (such as a keyboard), or two devices

with numeric keypad entry. In the first case, the display presents a 6-digit numeric code to the user, who then enters the code on the keypad. In the second case, the user of each device enters the same 6-digit number. Both of these cases provide MITM protection.

  - Out of band (OOB): This method uses an external means of communication, such as near-field communication (NFC) to exchange some information used in the pairing process. Pairing is completed using the Bluetooth radio, but requires information from the OOB mechanism. This provides only the level of MITM protection that is present in the OOB mechanism.

SSP is considered simple for the following reasons:

- In most cases, it does not require a user to generate a passkey.
- For use cases not requiring MITM protection, user interaction can be eliminated.
- For numeric comparison, MITM protection can be achieved with a simple equality comparison by the user.
- Using OOB with NFC enables pairing when devices simply get close, rather than requiring a lengthy discovery process.

Security Concerns

Prior to Bluetooth v2.1, encryption is not required and can be turned off at any time. Moreover, the encryption key is only good for approximately 23.5 hours; using a single encryption key longer than this time allows simple XOR attacks to retrieve the encryption key.

- Turning off encryption is required for several normal operations, so it is problematic to detect if encryption is disabled for a valid reason or for a security attack.

Bluetooth v2.1 addresses this in the following ways:

- Encryption is required for all non-SDP (Service Discovery Protocol) connections.
- A new Encryption Pause and Resume feature is used for all normal operations that require that encryption be disabled. This enables easy identification of normal operation from security attacks.
- The encryption key must be refreshed before it expires.

Link keys may be stored on the device file system, not on the Bluetooth chip itself. Many Bluetooth chip manufacturers let link keys be stored on the device—however, if the device is removable, this means that the link key moves with the device.

Security

Bluetooth implements confidentiality, authentication and key derivation with custom algorithms based on the SAFER+ block cipher. Bluetooth key generation is generally based on a Bluetooth PIN, which must be entered into both devices. This procedure might be modified if one of the devices has a fixed PIN (e.g., for headsets or similar devices with a restricted user interface). During pairing, an initialization key or master key is generated, using the E22 algorithm. The E0 stream cipher is used for encrypting packets, granting confidentiality, and is based on a shared cryptographic secret, namely a previously generated link key or master key. Those keys, used for subsequent encryption of data sent via the air interface, rely on the Bluetooth PIN, which has been entered into one or both devices.

An overview of Bluetooth vulnerabilities exploits was published in 2007 by Andreas Becker.

In September 2008, the National Institute of Standards and Technology (NIST) published a Guide to Bluetooth Security as a reference for organizations. It describes Bluetooth security capabilities and how to secure Bluetooth technologies effectively. While Bluetooth has its benefits, it is susceptible to denial-of-service attacks, eavesdropping, man-in-the-middle attacks, message modification, and resource misappropriation. Users and organizations must evaluate their acceptable level of risk and incorporate security into the lifecycle of Bluetooth devices. To help mitigate risks, included in the NIST document are security checklists with guidelines and recommendations for creating and maintaining secure Bluetooth piconets, headsets, and smart card readers.

Bluetooth v2.1 – finalized in 2007 with consumer devices first appearing in 2009 – makes significant changes to Bluetooth's security, including pairing. See the pairing mechanisms section for more about these changes.

Bluejacking

Bluejacking is the sending of either a picture or a message from one user to an unsuspecting user through Bluetooth wireless technology. Common applications include short messages, e.g., "You've just been bluejacked!" Bluejacking does not involve the removal or alteration of any data from the device. Bluejacking can also involve taking control of a mobile device wirelessly and phoning a premium rate line, owned by the bluejacker. Security advances have alleviated this issue.

Health Concerns

Bluetooth uses the microwave radio frequency spectrum in the 2.402 GHz to 2.480 GHz range, which is non-ionizing radiation, of similar bandwidth to the one used by wireless and mobile phones. No specific demonstration of harm has been demonstrated up to date, even if wireless transmission has been included by IARC in the possible carcinogen list. Maximum power output from a Bluetooth radio is 100 mW for class 1,

2.5 mW for class 2, and 1 mW for class 3 devices. Even the maximum power output of class 1 is a lower level than the lowest-powered mobile phones. UMTS and W-CDMA output 250 mW, GSM1800/1900 outputs 1000 mW, and GSM850/900 outputs 2000 mW.

Wi-Fi

Wi-Fi is a family of wireless networking technologies, based on the IEEE 802.11 family of standards, which are commonly used for local area networking of devices and Internet access. Wi–Fi is a trademark of the non-profit Wi-Fi Alliance, which restricts the use of the term Wi-Fi Certified to products that successfully complete interoperability certification testing. As of 2010, the Wi-Fi Alliance consisted of more than 375 companies from around the world. As of 2009, Wi-Fi-integrated circuit chips shipped approximately 580 million units yearly. Devices that can use Wi-Fi technologies include desktops and laptops, smartphones and tablets, smart TVs, printers, digital audio players, digital cameras, cars and drones.

Wi-Fi uses multiple parts of the IEEE 802 protocol family and is designed to seamlessly interwork with its wired sibling Ethernet. Compatible devices can network through a wireless access point to each other as well as to wired devices and the Internet. The different versions of Wi-Fi are specified by various IEEE 802.11 protocol standards, with the different radio technologies determining radio bands, and the maximum ranges, and speeds that may be achieved. Wi-Fi most commonly uses the 2.4 gigahertz (120 mm) UHF and 5 gigahertz (60 mm) SHF ISM radio bands; these bands are subdivided into multiple channels. Channels can be shared between networks but only one transmitter can locally transmit on a channel at any moment in time.

Wi-Fi's wavebands have relatively high absorption and work best for line-of-sight use. Many common obstructions such as walls, pillars, home appliances etc. may greatly reduce range, but this also helps minimize interference between different networks in crowded environments. An access point (or hotspot) often has a range of about 20 metres (66 feet) indoors while some modern access points claim up to a 150-metre (490-foot) range outdoors. Hotspot coverage can be as small as a single room with walls that block radio waves, or as large as many square kilometres using many overlapping access points with roaming permitted between them. Over time the speed and spectral efficiency of Wi-Fi has increased. As of 2019, at close range, some versions of Wi-Fi running on suitable hardware, can achieve speeds of over 1 Gbit/s (gigabit per second).

Wi-Fi is potentially more vulnerable to attack than wired networks because anyone within range of a network with a wireless network interface controller can attempt access. Therefore, to connect to a Wi-Fi network, a user typically needs the network name (the SSID) and a password. The password is used to encrypt Wi-Fi packets so as to block eavesdroppers. Wi-Fi Protected Access (WPA) is a family of technologies created to protect information moving across Wi-Fi networks and includes solutions for personal and enterprise networks. As the security landscape has changed over time security features of WPA have included stronger protections and new security practices.

Certification

The IEEE does not test equipment for compliance with their standards. The non-profit Wi-Fi Alliance was formed in 1999 to fill this void—to establish and enforce standards for interoperability and backward compatibility, and to promote wireless local-area-network technology. As of 2010, the Wi-Fi Alliance consists of more than 375 companies and includes 3Com (now owned by HPE/Hewlett-Packard Enterprise), Aironet (now owned by Cisco), Harris Semiconductor (now owned by Intersil), Lucent (now owned by Nokia), Nokia and Symbol Technologies (now owned by Zebra Technologies). The Wi-Fi Alliance enforces the use of the Wi-Fi brand to technologies based on the IEEE 802.11 standards from the IEEE. This includes wireless local area network (WLAN) connections, device to device connectivity (such as Wi-Fi Peer to Peer aka Wi-Fi Direct), Personal area network (PAN), local area network (LAN), and even some limited wide area network (WAN) connections. Manufacturers with membership in the Wi-Fi Alliance, whose products pass the certification process, gain the right to mark those products with the Wi-Fi logo.

Specifically, the certification process requires conformance to the IEEE 802.11 radio standards, the WPA and WPA2 security standards, and the EAP authentication standard. Certification may optionally include tests of IEEE 802.11 draft standards, interaction with cellular-phone technology in converged devices, and features relating to security set-up, multimedia, and power-saving.

Not every Wi-Fi device is submitted for certification. The lack of Wi-Fi certification does not necessarily imply that a device is incompatible with other Wi-Fi devices. The Wi-Fi Alliance may or may not sanction derivative terms, such as Super Wi-Fi, coined by the US Federal Communications Commission (FCC) to describe proposed networking in the UHF TV band in the US.

Versions

Generation/IEEE Standard	Maximum Linkrate	Adopted	Frequency
Wi-Fi 6 (802.11ax)	600–9608 Mbit/s	2019	2.4/5 GHz 1–6 GHz ISM
Wi-Fi 5 (802.11ac)	433–6933 Mbit/s	2014	5 GHz
Wi-Fi 4 (802.11n)	72–600 Mbit/s	2009	2.4/5 GHz

802.11g	3–54 Mbit/s	2003	2.4 GHz
802.11a	1.5 to 54 Mbit/s	1999	5 GHz
802.11b	1 to 11 Mbit/s	1999	
(Wi-Fi 1, Wi-Fi 2, Wi-Fi 3 are unbranded but have unofficial assignments.)			

Equipment frequently support multiple versions of Wi-Fi. To communicate, devices must use a common Wi-Fi version. The versions differ between the radio wavebands they operate on, the radio bandwidth they occupy, the maximum data rates they can support and other details. Some versions permit the use of multiple antennas, which permits greater speeds as well as reduced interference.

Historically, equipment has simply listed the versions of Wi-Fi using the name of the IEEE standard that it supports. In 2018, the Wi-Fi alliance standardized generational numbering so that equipment can indicate that it supports Wi-Fi 4 (if the equipment supports 802.11n), Wi-Fi 5 (802.11ac) and Wi-Fi 6 (802.11ax). These generations have a high degree of backward compatibility with previous versions. The alliances have stated that the generational level 4, 5, or 6 can be indicated in the user interface when connected, along with the signal strength.

The full list of versions of Wi-Fi is: 802.11a, 802.11b, 802.11g, 802.11n (Wi-Fi 4), 802.11h, 802.11i, 802.11-2007, 802.11-2012, 802.11ac (Wi-Fi 5), 802.11ad, 802.11af, 802.11-2016, 802.11ah, 802.11ai, 802.11aj, 802.11aq, 802.11ax (Wi-Fi 6), 802.11ay.

Uses of WiFi

Internet Access

Wi-Fi technology may be used to provide local network and Internet access to devices that are within Wi-Fi range of one or more routers that are connected to the Internet. The coverage of one or more interconnected access points (hotspots) can extend from an area as small as a few rooms to as large as many square kilometres. Coverage in the larger area may require a group of access points with overlapping coverage. For example, public outdoor Wi-Fi technology has been used successfully in wireless mesh networks in London. An international example is Fon.

Wi-Fi provides service in private homes, businesses, as well as in public spaces. Wi-Fi hotspots may be set up either free-of-charge or commercially, often using a captive portal webpage for access. Organizations, enthusiasts, authorities and businesses, such as airports, hotels, and restaurants, often provide free or paid-use hotspots to attract customers, to provide services to promote business in selected areas.

Routers often incorporate a digital subscriber line modem or a cable modem and a Wi-Fi access point, are frequently set up in homes and other buildings, to provide Internet access and internetworking for the structure.

Similarly, battery-powered routers may include a cellular Internet radio modem and Wi-Fi access point. When subscribed to a cellular data carrier, they allow nearby Wi-Fi stations to access the Internet over 2G, 3G, or 4G networks using the tethering technique. Many smartphones have a built-in capability of this sort, including those based on Android, BlackBerry, Bada, iOS (iPhone), Windows Phone, and Symbian, though carriers often disable the feature, or charge a separate fee to enable it, especially for customers with unlimited data plans. "Internet packs" provide standalone facilities of this type as well, without use of a smartphone; examples include the MiFi- and WiBro-branded devices. Some laptops that have a cellular modem card can also act as mobile Internet Wi-Fi access points.

Many traditional university campuses in the developed world provide at least partial Wi-Fi coverage. Carnegie Mellon University built the first campus-wide wireless Internet network, called Wireless Andrew, at its Pittsburgh campus in 1993 before Wi-Fi branding originated. By February 1997, the CMU Wi-Fi zone was fully operational. Many universities collaborate in providing Wi-Fi access to students and staff through the Eduroam international authentication infrastructure.

City-wide

An outdoor Wi-Fi access point.

In the early 2000s, many cities around the world announced plans to construct city-wide Wi-Fi networks. There are many successful examples; in 2004, Mysore (Mysuru) became India's first Wi-Fi-enabled city. A company called WiFiyNet has set up hotspots in Mysore, covering the complete city and a few nearby villages.

In 2005, St. Cloud, Florida and Sunnyvale, California, became the first cities in the United States to offer citywide free Wi-Fi (from MetroFi). Minneapolis has generated $1.2 million in profit annually for its provider.

In May 2010, London mayor Boris Johnson pledged to have London-wide Wi-Fi by 2012. Several boroughs including Westminster and Islington already had extensive outdoor Wi-Fi coverage at that point.

Officials in South Korea's capital Seoul are moving to provide free Internet access at more than 10,000 locations around the city, including outdoor public spaces, major streets and densely populated residential areas. Seoul will grant leases to KT, LG Telecom, and SK Telecom. The companies will invest $44 million in the project, which was to be completed in 2015.

Geolocation

Wi-Fi positioning systems use the positions of Wi-Fi hotspots to identify a device's location.

Operational Principles

Wi-Fi stations communicate by sending each other data packets: blocks of data individually sent and delivered over radio. As with all radio, this is done by the modulating and demodulation of carrier waves. Different versions of Wi-Fi use different techniques, 802.11b uses DSSS on a single carrier, whereas 802.11a, Wi-Fi 4, 5 and 6 use multiple carriers on slightly different frequencies within the channel (OFDM).

As with other IEEE 802 LANs, stations come programmed with a globally unique 48-bit MAC address (often printed on the equipment) so that each Wi-Fi station has a unique address. The MAC addresses are used to specify both the destination and the source of each data packet. Wi-Fi establishes link-level connections, which can be defined using both the destination and source addresses. On reception of a transmission, the receiver uses the destination address to determine whether the transmission is relevant to the station or should be ignored. A network interface normally does not accept packets addressed to other Wi-Fi stations.

Due to the ubiquity of Wi-Fi and the ever-decreasing cost of the hardware needed to support it, most manufacturers now build Wi-Fi interfaces directly into PC motherboards, eliminating the need for installation of a separate network card.

Channels are used half duplex and can be time-shared by multiple networks. When communication happens on the same channel, any information sent by one computer is locally received by all, even if that information is intended for just one destination. The network interface card interrupts the CPU only when applicable packets are received: the card ignores information not addressed to it. Use of the same channel also means that the data bandwidth is shared, such that, for example, available data bandwidth to each device is halved when two stations are actively transmitting.

A collision happens when two stations attempt to transmit at the same time. They corrupt transmitted data and require stations to re-transmit. The lost data and re-transmission reduces throughput. In the worst case, where multiple active hosts connected with maximum allowed cable length attempt to transmit many short frames, excessive collisions can reduce throughput dramatically. A scheme known as carrier sense multiple access with collision avoidance (CSMA/CA) governs the way the computers share the channel.

Waveband

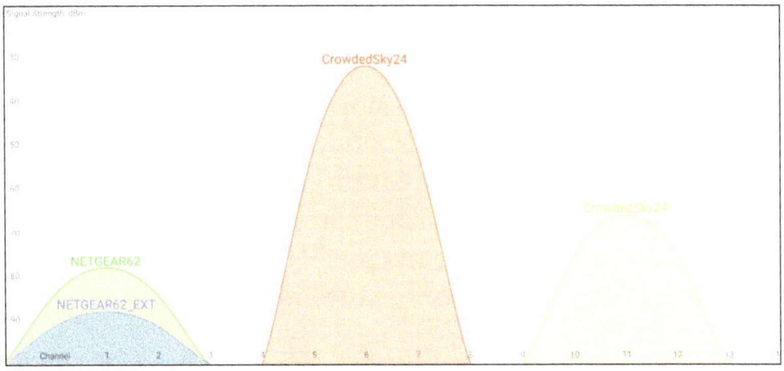

In the 2.4 GHz wavebands as well as others, transmitters straddle multiple channels. Overlapping channels can suffer from interference unless this is a small portion of the total received power.

A keychain-size Wi-Fi detector.

The 802.11 standard provides several distinct radio frequency ranges for use in Wi-Fi communications: 900 MHz, 2.4 GHz, 5 GHz, 5.9 GHz, and 60 GHz bands. Each range is divided into a multitude of channels. Countries apply their own regulations to the allowable channels, allowed users and maximum power levels within these frequency ranges. The ISM band ranges are also often used.

802.11b/g/n can use the 2.4 GHz ISM band, operating in the United States under Part 15 Rules and Regulations. In this frequency band equipment may occasionally suffer interference from microwave ovens, cordless telephones, USB 3.0 hubs, and Bluetooth devices.

Spectrum assignments and operational limitations are not consistent worldwide: Australia and Europe allow for an additional two channels (12, 13) beyond the 11 permitted in the United States for the 2.4 GHz band, while Japan has three more (12–14). In the US and other countries, 802.11a and 802.11g devices may be operated without a license, as allowed in Part 15 of the FCC Rules and Regulations.

802.11a/h/j/n/ac/ax can use the 5 GHz U-NII band, which, for much of the world, offers at least 23 non-overlapping 20 MHz channels rather than the 2.4 GHz ISM frequency band, where the channels are only 5 MHz wide. In general, lower frequencies have better range but have less capacity. The 5 GHz bands are absorbed to a greater degree by common building materials than the 2.4 GHz bands, and usually give shorter range.

As 802.11 specifications evolved to support higher throughput, the protocols have become much more efficient in their use of bandwidth. Additionally they have gained the ability to aggregate (or 'bond') channels together to gain still more throughput where the bandwidth is available. 802.11n allows for double radio spectrum/bandwidth (40 MHz- 8 channels) compared to 802.11a or 802.11g (20 MHz). 802.11n can also be set to limit itself to 20 MHz bandwidth to prevent interference in dense communities. In the 5 GHz band, 20 MHz, 40 MHz, 80 MHz, and 160 MHz bandwidth signals are permitted with some restrictions, giving much faster connections.

Communication Stack

Wi-Fi is part of the IEEE 802 protocol family. The data is organized into 802.11 frames that are very similar to Ethernet frames at the data link layer, but with extra address fields. MAC addresses are used as network addresses for routing over the LAN.

Wi-Fi's MAC and physical layer (PHY) specifications are defined by IEEE 802.11 for modulating and receiving one or more carrier waves to transmit the data in the infrared, and 2.4, 3.6, 5, or 60 GHz frequency bands. They are created and maintained by the IEEE LAN/MAN Standards Committee (IEEE 802). The base version of the standard was released in 1997, and has had many subsequent amendments. The standard and amendments provide the basis for wireless network products using the Wi-Fi brand. While each amendment is officially revoked when it is incorporated in the latest version of the standard, the corporate world tends to market to the revisions because they concisely denote capabilities of their products. As a result, in the market place, each revision tends to become its own standard.

In addition to 802.11 the IEEE 802 protocol family has specific provisions for Wi-Fi. These are required because Ethernet's cable-based media are not usually shared, whereas with wireless all transmissions are received by all stations within range that employ that radio channel. While Ethernet has essentially negligible error rates, wireless communication media are subject to significant interference. Therefore, accurate transmission is not guaranteed so delivery is therefore a best-effort delivery mechanism. Because of this, for Wi-Fi, the Logical Link Control (LLC) specified by IEEE 802.2 employs Wi-Fi's media access control (MAC) protocols to manage retries without relying on higher levels of the protocol stack.

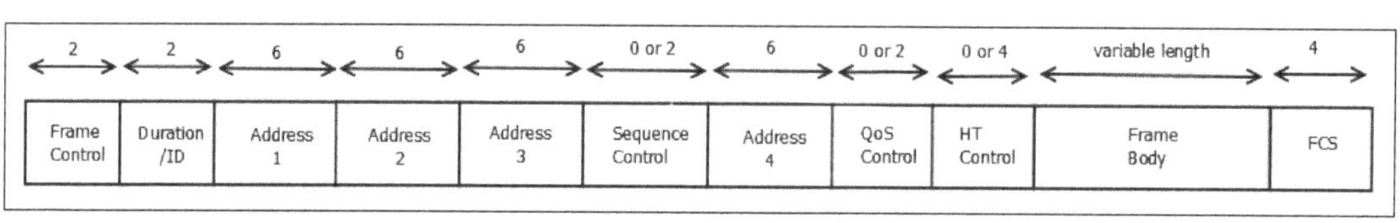

For internetworking purposes Wi-Fi is usually layered as a link layer (equivalent to the physical and data link layers of the OSI model) below the internet layer of the Internet Protocol. This means that nodes have an associated internet address and, with suitable connectivity, this allows full Internet access.

Modes

Infrastructure

In infrastructure mode, which is the most common mode used, all communications goes through a base station. For communications within the network, this introduces an extra use of the airwaves, but has the advantage that any two stations that can communicate with the base station can also communicate through the base station, which enormously simplifies the protocols.

Ad Hoc and Wi-Fi Direct

Wi-Fi also allows communications directly from one computer to another without an access point intermediary. This is called ad hoc Wi-Fi transmission. Different types of ad hoc network exist. In the simplest case network nodes must talk directly to each other. In more complex protocols nodes may forward packets, and nodes keep track of how to reach other nodes, even if they move around.

Ad-hoc mode was first invented and realized by Chai Keong Toh in his 1996 invention of Wi-Fi ad-hoc routing, implemented on Lucent WaveLAN 802.11a wireless on IBM ThinkPads over a size nodes scenario spanning a region of over a mile. The success was recorded in Mobile Computing magazine and later published formally in IEEE Transactions on Wireless Communications, 2002 and ACM SIGMETRICS Performance Evaluation Review, 2001.

This wireless ad hoc network mode has proven popular with multiplayer handheld game consoles, such as the Nintendo DS, PlayStation Portable, digital cameras, and other consumer electronics devices. Some devices can also share their Internet connection using ad hoc, becoming hotspots or "virtual routers".

Similarly, the Wi-Fi Alliance promotes the specification Wi-Fi Direct for file transfers and media sharing through a new discovery- and security-methodology. Wi-Fi Direct launched in October 2010.

Another mode of direct communication over Wi-Fi is Tunneled Direct Link Setup (TDLS), which enables two devices on the same Wi-Fi network to communicate directly, instead of via the access point.

Multiple Access Points

An Extended Service Set may be formed by deploying multiple access points that are configured with the same SSID and security settings. Wi-Fi client devices typically connect to the access point that can provide the strongest signal within that service set.

Increasing the number of Wi-Fi access points for a network provides redundancy, better range, support for fast roaming and increased overall network-capacity by using more channels or by defining smaller cells. Except for the smallest implementations (such as home or small office networks), Wi-Fi implementations have moved toward "thin" access points, with more of the network intelligence housed in a centralized network appliance, relegating individual access points to the role of "dumb" transceivers. Outdoor applications may use mesh topologies.

Performance

Parabolic dishes transmit and receive the radio waves only in particular directions and can give much greater range than omnidirection antennas.

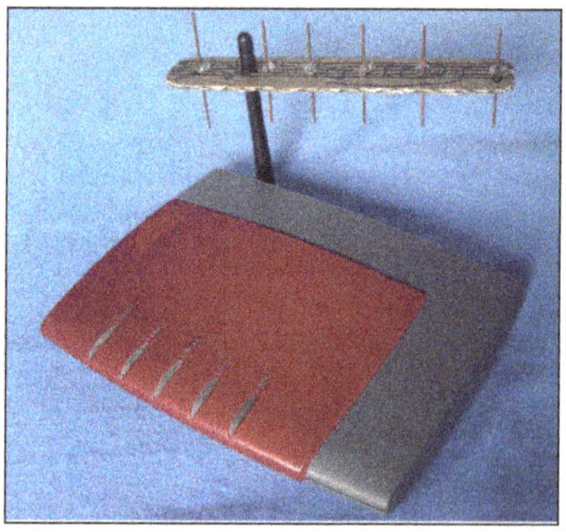

Yagi-Uda antennas, widely used for television reception, are relatively compact at Wi-Fi wavelengths.

Wi-Fi operational range depends on factors such as the frequency band, radio power output, receiver sensitivity, antenna gain and antenna type as well as the modulation technique. In addition, propagation characteristics of the signals can have a big impact.

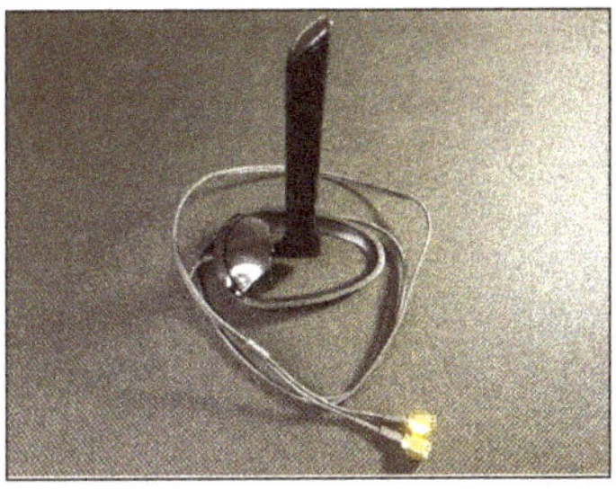

Antenna of wireless network interface controller Gigabyte GC-WB867D-I.
Simple stick-like antennas like these have unidirectional reception
and relatively low range of 20m or so.

At longer distances, and with greater signal absorption, speed is usually reduced.

Transmitter Power

Compared to cell phones and similar technology, Wi-Fi transmitters are low power devices. In general, the maximum amount of power that a Wi-Fi device can transmit is limited by local regulations, such as FCC Part 15 in the US. Equivalent isotropically radiated power (EIRP) in the European Union is limited to 20 dBm (100 mW).

To reach requirements for wireless LAN applications, Wi-Fi has higher power consumption compared to some other standards designed to support wireless personal area network (PAN) applications. For example, Bluetooth provides a much shorter propagation range between 1 and 100m and so in general have a lower power consumption. Other low-power technologies such as ZigBee have fairly long range, but much lower data rate. The high power consumption of Wi-Fi makes battery life in some mobile devices a concern.

Antenna

An access point compliant with either 802.11b or 802.11g, using the stock omnidirectional antenna might have a range of 100 m (0.062 mi). The same radio with an external semi parabolic antenna (15 dB gain) with a similarly equipped receiver at the far end might have a range over 20 miles.

Higher gain rating (dBi) indicates further deviation (generally toward the horizontal) from a theoretical, perfect isotropic radiator, and therefore the antenna can project or accept a usable signal further in particular directions, as compared to a similar output power on a more isotropic antenna. For example, an 8 dBi antenna used with a 100 mW driver has a similar horizontal range to a 6 dBi antenna being driven at 500 mW. Note that this assumes that radiation in the vertical is lost; this may not be the case in some situations, especially in large buildings or within a waveguide. In the above example, a directional waveguide could cause the low power 6 dBi antenna to project much further in a single direction than the 8 dBi antenna, which is not in a waveguide, even if they are both driven at 100 mW.

On wireless routers with detachable antennas, it is possible to improve range by fitting upgraded antennas that provide higher gain in particular directions. Outdoor ranges can be improved to many kilometres through the use of high gain directional antennas at the router and remote device(s).

MIMO (Multiple-input and Multiple-output)

Wi-Fi 4 and higher standards allow devices to have multiple antennas on transmitters and receivers. Multiple antennas enable the equipment to exploit multipath propagation on the same frequency bands giving much faster speeds and greater range.

Wi-Fi 4 can more than double the range over previous standards. The Wi-Fi 5 standard uses the 5 GHz band exclusively and is capable of multi-station WLAN throughput of at least 1 gigabit per second, and a single station throughput of at least 500 Mbit/s. In the first quarter of 2016, The Wi-Fi Alliance certifies devices compliant with the 802.11ac standard as "Wi-Fi CERTIFIED ac". This standard uses several signal processing techniques such as multi-user MIMO and 4X4 Spatial Multiplexing streams, and large channel bandwidth (160 MHz) to achieve the Gigabit throughput. According to a study by IHS Technology, 70% of all access point sales revenue In the first quarter of 2016 came from 802.11ac devices.

Radio Propagation

With Wi-Fi signals line-of-sight usually works best, but signals can transmit, absorb, reflect, and diffract through and around structures, both man made, and natural.

This Netgear Wi-Fi router contains dual bands for transmitting the 802.11 standard across the 2.4 and 5 GHz spectrums and supports MIMO.

Due to the complex nature of radio propagation at typical Wi-Fi frequencies, particularly around trees and buildings, algorithms can only approximately predict Wi-Fi signal strength for any given area in relation to a transmitter. This effect does not apply equally to long-range Wi-Fi, since longer links typically operate from towers that transmit above the surrounding foliage.

Mobile use of Wi-Fi over wider ranges is limited, for instance, to uses such as in an automobile moving from one hotspot to another. Other wireless technologies are more suitable for communicating with moving vehicles.

Distance Records

Distance records (using non-standard devices) include 382 km (237 mi) in June 2007, held by Ermanno Pietrosemoli and EsLaRed of Venezuela, transferring about 3 MB of data between the mountain-tops of El Águila and Platillon. The Swedish Space Agency transferred data 420 km (260 mi), using 6 watt amplifiers to reach an overhead stratospheric balloon.

Interference

Wi-Fi connections can be blocked or the Internet speed lowered by having other devices in the same area. Wi-Fi protocols are designed to share the wavebands reasonably fairly, and this often works with little to no disruption. To minimize collisions with Wi-Fi and non-Wi-Fi devices, Wi-Fi employs Carrier-sense multiple access with collision avoidance (CSMA/CA), where transmitters listen before transmitting, and delay transmission of packets if they detect that other devices are active on the channel, or if noise is detected from adjacent channels or from non-Wi-Fi sources. Nevertheless, Wi-Fi networks are still susceptible to the hidden node and exposed node problem.

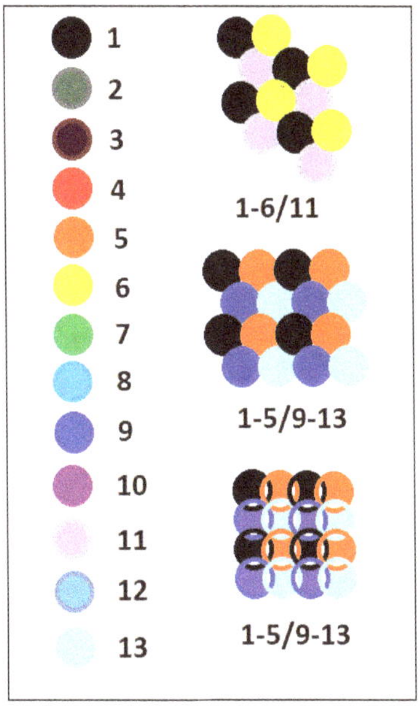

Network planning frequency allocations for North America and Europe. Using these types of frequency allocations can help minimize co-channel and adjacent-channel interference.

A standard speed Wi-Fi signal occupies five channels in the 2.4 GHz band. Interference can be caused by overlapping channels. Any two channel numbers that differ by five or more, such as 2 and 7, do not overlap (no adjacent-channel interference). The oft-repeated adage that channels 1, 6, and 11 are the only non-overlapping channels is, therefore, not accurate. Channels 1, 6, and 11 are the only group of three non-overlapping channels in North America. However, whether the overlap is significant depends on physical spacing. Channels that are four apart interfere a negligible amount-much less than reusing channels (which causes co-channel interference)-if transmitters are at least a few metres apart. In Europe and Japan where channel 13 is available, using Channels 1, 5, 9, and 13 for 802.11g and 802.11n is recommended.

However, many 2.4 GHz 802.11b and 802.11g access-points default to the same channel on initial startup, contributing to congestion on certain channels. Wi-Fi pollution, or an excessive number of access points in the area, can prevent access and interfere with other devices' use of other access points as well as with decreased signal-to-noise ratio (SNR) between access points. These issues can become a problem in high-density areas, such as large apartment complexes or office buildings with many Wi-Fi access points. Wi-Fi 6 has greatly improved power control, and suffers less from interference in congested areas.

Other devices use the 2.4 GHz band: microwave ovens, ISM band devices, security cameras, ZigBee devices, Bluetooth devices, video senders, cordless phones, baby monitors, and, in some countries, amateur radio, all of which can cause significant additional interference. It is also an issue when municipalities or other large entities (such as universities) seek to provide large area coverage. On some 5 GHz bands interference from radar systems can occur in some places. For base stations that support those bands they employ Dynamic Frequency Selection which listens for radar, and if it is found, will not permit a network on that band.

These bands can be used by low power transmitters without a license, and with few restrictions. However, while unintended interference is common, users that have been found to cause deliberate interference (particularly for attempting to locally monopolize these bands for commercial purposes) have been issued large fines.

Throughput

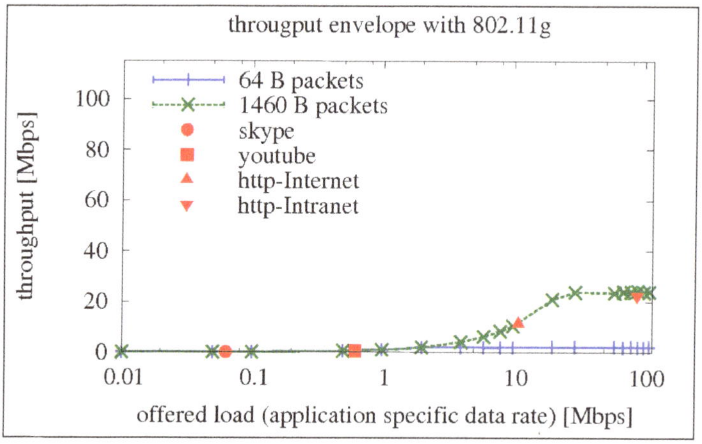

Graphical representation of Wi-Fi application specific (UDP) performance envelope 2.4 GHz band, with 802.11g.

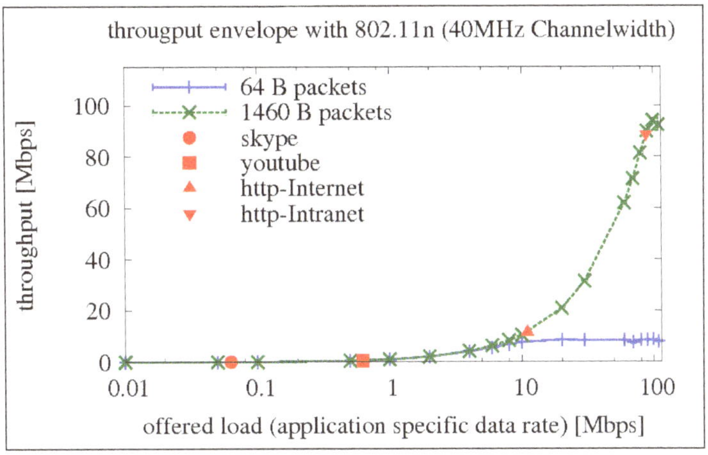

Graphical representation of Wi-Fi application specific (UDP) performance envelope 2.4 GHz band, with 802.11n with 40 MHz.

Various layer 2 variants of IEEE 802.11 have different characteristics. Across all flavours of 802.11, maximum achievable throughputs are either given based on measurements under ideal conditions or in the layer 2 data rates. This, however, does not apply to typical deployments in which data are transferred between two endpoints of which at least one is typically connected to a wired infrastructure, and the other is connected to an infrastructure via a wireless link.

This means that typically data frames pass an 802.11 (WLAN) medium and are being converted to 802.3 (Ethernet) or vice versa.

Due to the difference in the frame (header) lengths of these two media, the packet size of an application determines the speed of the data transfer. This means that an application that uses small packets (e.g., VoIP) creates a data flow with a high overhead traffic (low goodput).

Other factors that contribute to the overall application data rate are the speed with which the application transmits the packets (i.e., the data rate) and the energy with which the wireless signal is received. The latter is determined by distance and by the configured output power of the communicating devices.

The same references apply to the attached throughput graphs, which show measurements of UDP throughput measurements. Each represents an average throughput of 25 measurements (the error bars are there, but barely visible due to the small variation), is with a specific packet size (small or large), and with a specific data rate (10 kbit/s – 100 Mbit/s). Markers for traffic profiles of common applications are included as well. This text and measurements do not cover packet errors but information about this can be found at the above references. The table below shows the maximum achievable (application specific) UDP throughput in the same scenarios (same references again) with various different WLAN (802.11) flavours. The measurement hosts have been 25 metres apart from each other; loss is again ignored.

Hardware

An embedded RouterBoard 112 with U.FL-RSMA pigtail and R52 mini PCI Wi-Fi card widely used by wireless Internet service providers (WISPs) in the Czech Republic.

OSBRiDGE 3GN – 802.11n Access Point and UMTS/GSM Gateway in one device.

Wi-Fi allows wireless deployment of local area networks (LANs). Also, spaces where cables cannot be run, such as outdoor areas and historical buildings, can host wireless LANs. However, building walls of certain materials, such as stone with high metal content, can block Wi-Fi signals.

A Wi-Fi device is a short-range wireless device. Wi-Fi devices are fabricated on RF CMOS integrated circuit (RF circuit) chips.

Since the early 2000s, manufacturers are building wireless network adapters into most laptops. The price of chipsets for Wi-Fi continues to drop, making it an economical networking option included in ever more devices.

Different competitive brands of access points and client network-interfaces can inter-operate at a basic level of service. Products designated as "Wi-Fi Certified" by the Wi-Fi Alliance are backward compatible. Unlike mobile phones, any standard Wi-Fi device works anywhere in the world.

Access Point

An AirPort wireless G Wi-Fi adapter from an Apple MacBook.

A wireless access point (WAP) connects a group of wireless devices to an adjacent wired LAN. An access point resembles a network hub, relaying data between connected wireless devices in addition to a (usually) single connected wired device, most often an Ethernet hub or switch, allowing wireless devices to communicate with other wired devices.

Wireless Adapter

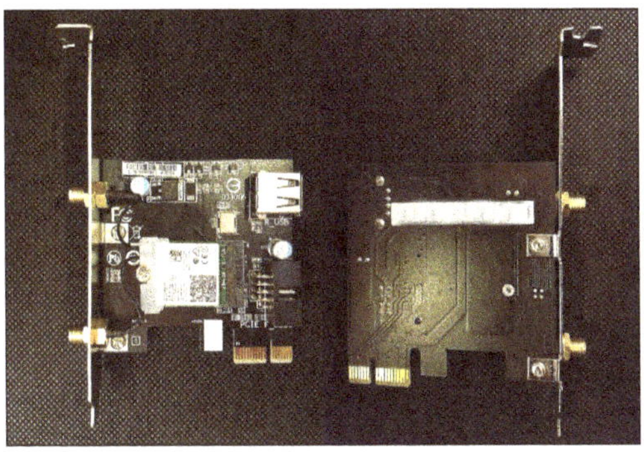

Wireless network interface controller Gigabyte GC-WB867D-I.

Wireless adapters allow devices to connect to a wireless network. These adapters connect to devices using various external or internal interconnects such as PCI, miniPCI, USB, ExpressCard, Cardbus, and PC Card. As of 2010, most newer laptop computers come equipped with built in internal adapters.

Router

Wireless routers integrate a Wireless Access Point, Ethernet switch, and internal router firmware application that provides IP routing, NAT, and DNS forwarding through an integrated WAN-interface. A wireless router allows wired and wireless Ethernet LAN devices to connect to a (usually) single WAN device such as a cable modem, DSL modem or optical modem. A wireless router allows all three devices, mainly the access point and router, to be configured through one central utility. This utility is usually an integrated web server that is accessible to wired and wireless LAN clients and often optionally to WAN clients. This utility may also be an application that is run on a computer, as is the case with as Apple's AirPort, which is managed with the AirPort Utility on macOS and iOS.

Bridge

Wireless network bridges can act to connect two networks to form a single network at the data-link layer over Wi-Fi. The main standard is the wireless distribution system (WDS).

Wireless bridging can connect a wired network to a wireless network. A bridge differs from an access point: an access point typically connects wireless devices to one wired network. Two wireless bridge devices may be used to connect two wired networks over a wireless link, useful in situations where a wired connection may be unavailable, such as between two separate homes or for devices that have no wireless networking capability (but have wired networking capability), such as consumer entertainment devices; alternatively, a wireless bridge can be used to enable a device that supports a wired connection to operate at a wireless networking standard that is faster than supported by the wireless network connectivity feature (external dongle or inbuilt) supported by the device (e.g., enabling Wireless-N speeds (up to the maximum supported speed on the wired Ethernet port on both the bridge and connected devices including the wireless access point) for a device that only supports Wireless-G). A dual-band wireless bridge can also be used to enable 5 GHz wireless network operation on a device that only supports 2.4 GHz wireless and has a wired Ethernet port.

Wireless range-extenders or wireless repeaters can extend the range of an existing wireless network. Strategically placed range-extenders can elongate a signal area or allow for the signal area to reach around barriers such as those pertaining in L-shaped corridors. Wireless devices connected through repeaters suffer from an increased latency for each hop, and there may be a reduction in the maximum available data throughput. In addition, the effect of additional users using a network employing wireless range-extenders is to consume the available bandwidth faster than would be the case whereby a single user migrates around a network employing extenders. For this reason, wireless range-extenders work best in networks supporting low traffic throughput requirements, such as for cases whereby a single user with a Wi-Fi equipped tablet migrates around the combined extended and non-extended portions of the total connected network. Also, a wireless device connected to any of the repeaters in the chain has data throughput limited by the "weakest link" in the chain between the connection origin and connection end. Networks using wireless extenders are more prone to degradation from interference from neighboring access points that border portions of the extended network and that happen to occupy the same channel as the extended network.

Embedded Systems

Embedded Serial-to-Wi-Fi module.

The security standard, Wi-Fi Protected Setup, allows embedded devices with limited graphical user interface to connect to the Internet with ease. Wi-Fi Protected Setup has 2 configurations: The Push Button configuration and the PIN configuration. These embedded devices are also called The Internet of Things and are low-power, battery-operated embedded systems. A number of Wi-Fi manufacturers design chips and modules for embedded Wi-Fi, such as GainSpan.

Increasingly in the last few years, embedded Wi-Fi modules have become available that incorporate a real-time operating system and provide a simple means of wirelessly enabling any device that can communicate via a serial port. This allows the design of simple monitoring devices. An example is a portable ECG device monitoring a patient at home. This Wi-Fi-enabled device can communicate via the Internet.

These Wi-Fi modules are designed by OEMs so that implementers need only minimal Wi-Fi knowledge to provide Wi-Fi connectivity for their products.

In June 2014, Texas Instruments introduced the first ARM Cortex-M4 microcontroller with an onboard dedicated Wi-Fi MCU, the SimpleLink CC3200. It makes embedded systems with Wi-Fi connectivity possible to build as single-chip devices, which reduces their cost and minimum size, making it more practical to build wireless-networked controllers into inexpensive ordinary objects.

Network Security

The main issue with wireless network security is its simplified access to the network compared to traditional wired networks such as Ethernet. With wired networking, one must either gain access to a building (physically connecting into the internal network), or break through an external firewall. To access Wi-Fi, one must merely be within the range of the Wi-Fi network. Most business networks protect sensitive data and systems by attempting to disallow external access. Enabling wireless connectivity reduces security if the network uses inadequate or no encryption.

An attacker who has gained access to a Wi-Fi network router can initiate a DNS spoofing attack against any other user of the network by forging a response before the queried DNS server has a chance to reply.

Securing Methods

A common measure to deter unauthorized users involves hiding the access point's name by disabling the SSID broadcast. While effective against the casual user, it is ineffective as a security method because the SSID is broadcast in the clear in response to a client SSID query. Another method is to only allow computers with known MAC addresses to join the network, but determined eavesdroppers may be able to join the network by spoofing an authorized address.

Wired Equivalent Privacy (WEP) encryption was designed to protect against casual snooping but it is no longer considered secure. Tools such as AirSnort or Aircrack-ng can quickly recover WEP encryption keys. Because of WEP's weakness the Wi-Fi Alliance approved Wi-Fi Protected Access (WPA) which uses TKIP. WPA was specifically designed to work with older equipment usually through a firmware upgrade. Though more secure than WEP, WPA has known vulnerabilities.

The more secure WPA2 using Advanced Encryption Standard was introduced in 2004 and is supported by most new Wi-Fi devices. WPA2 is fully compatible with WPA. In 2017, a flaw in the WPA2 protocol was discovered, allowing a key replay attack, known as KRACK.

A flaw in a feature added to Wi-Fi in 2007; called Wi-Fi Protected Setup (WPS), let WPA and WPA2 security be bypassed, and effectively broken in many situations. The only remedy as of late 2011 was to turn off Wi-Fi Protected Setup, which is not always possible.

Virtual Private Networks can be used to improve the confidentiality of data carried through Wi-Fi networks, especially public Wi-Fi networks.

Data Security Risks

The older wireless encryption-standard, Wired Equivalent Privacy (WEP), has been shown easily breakable even when correctly configured. Wi-Fi Protected Access (WPA and WPA2) encryption, which became available in devices in 2003, aimed to solve this problem. Wi-Fi access points typically default to an encryption-free (open) mode. Novice users benefit from a zero-configuration device that works out-of-the-box, but this default does not enable any wireless security, providing open wireless access to a LAN. To turn security on requires the user to configure the device, usually via a software graphical user interface (GUI). On unencrypted Wi-Fi networks connecting devices can monitor and record data (including personal information). Such networks can only be secured by using other means of protection, such as a VPN or secure Hypertext Transfer Protocol over Transport Layer Security (HTTPS).

Wi-Fi Protected Access encryption (WPA2) is considered secure, provided a strong passphrase is used. In 2018, WPA3 was announced as a replacement for WPA2, increasing security; it rolled out on June 26.

Piggybacking

Piggybacking refers to access to a wireless Internet connection by bringing one's own computer within the range of another's wireless connection, and using that service without the subscriber's explicit permission or knowledge.

During the early popular adoption of 802.11, providing open access points for anyone within range to use was encouraged to cultivate wireless community networks, particularly since people on average use only a fraction of their downstream bandwidth at any given time.

Recreational logging and mapping of other people's access points has become known as wardriving. Indeed, many access points are intentionally installed without security turned on so that they can be used as a free service. Providing access to one's Internet connection in this fashion may breach the Terms of Service or contract with the ISP. These activities do not result in sanctions in most jurisdictions; however, legislation and case law differ considerably across the world. A proposal to leave graffiti describing available services was called warchalking.

Piggybacking often occurs unintentionally – a technically unfamiliar user might not change the default "unsecured" settings to their access point and operating systems can be configured to connect automatically to any available wireless network. A user who happens to start up a laptop in the vicinity of an access point may find the computer has joined the network without any visible indication. Moreover, a user intending to join one network may instead end up on another one if the latter has a stronger signal. In combination with automatic discovery of other network resources this could possibly lead wireless users to send sensitive data to the wrong middle-man when seeking a destination. For example, a user could inadvertently use an unsecure network to log into a website, thereby making the login credentials available to anyone listening, if the website uses an unsecure protocol such as plain HTTP without TLS.

An unauthorized user can obtain security information (factory preset passphrase and/or Wi-Fi Protected Setup PIN) from a label on a wireless access point can use this information (or connect by the Wi-Fi Protected Setup pushbutton method) to commit unauthorized and/or unlawful activities.

References

- Wireless-transmission, data-communication-computer-network: tutorialspoint.com, Retrieved 27 July, 2020
- IEEE Std 802.15.1–2005 – IEEE Standard for Information technology – Telecommunications and information exchange between systems – Local and metropolitan area networks – Specific requirements Part 15.1: Wireless Medium Access Control (MAC) and Physical Layer (PHY)

Specifications for Wireless Personal Area Networks (W Pans). Ieeexplore.ieee.org. doi:10.1109/IEEESTD.2005.96290. ISBN 978-0-7381-4708-6. Retrieved 4 September 2010

- Lemstra, Wolter; Hayes, Vic; Groenewegen, John (2010). The Innovation Journey of Wi-Fi: The Road to Global Success. Cambridge University Press. p. 121. ISBN 978-0-521-19971-1. Archived from the original on 12 November 2012. Retrieved 6 October 2011
- Infrared-Transmission: tutorialspoint.com, , Retrieved 14 February, 2020
- Rudersdorfer, Ralf (2013). Radio Receiver Technology: Principles, Architectures and Applications. John Wiley and Sons. ISBN 978-1118647844

4

Analog Communication and Modulation

Analog communication uses continuous signals and modulation for transmission of data such as voice, image, video, etc. Some of its concepts are amplitude modulation, single-sideband modulation, space modulation, angle modulation, etc. All these concepts related to analog communication have been carefully analyzed in this chapter.

A continuous time varying signal, which represents a time varying quantity can be termed as an Analog Signal. This signal keeps on varying with respect to time, according to the instantaneous values of the quantity, which represents it.

Let us consider a tap that fills a tank of 100 liters capacity in an hour (6 AM to 7 AM). The portion of filling the tank is varied by the varying time. Which means, after 15 minutes (6:15 AM) the quarter portion of the tank gets filled, whereas at 6:45 AM, 3/4th of the tank is filled.

If we try to plot the varying portions of water in the tank according to the varying time, it would look like the following figure.

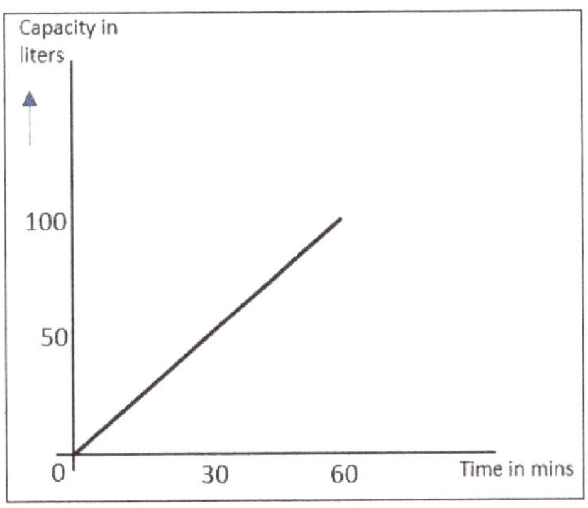

As the result shown in this image varies (increases) according to time, this time varying quantity can be understood as Analog quantity. The signal which represents this condition with an inclined line in the figure, is an Analog Signal. The communication based on analog signals and analog values is called as Analog Communication.

Analog Wireless Technology

In analog systems, signals of varying frequency or amplitude are used to modulate the carrier waves. Analog signals are continuously changing (infinite values) and are represented as a series of sine waves. The AM and FM radio transmissions are the most common examples of analog transmission.

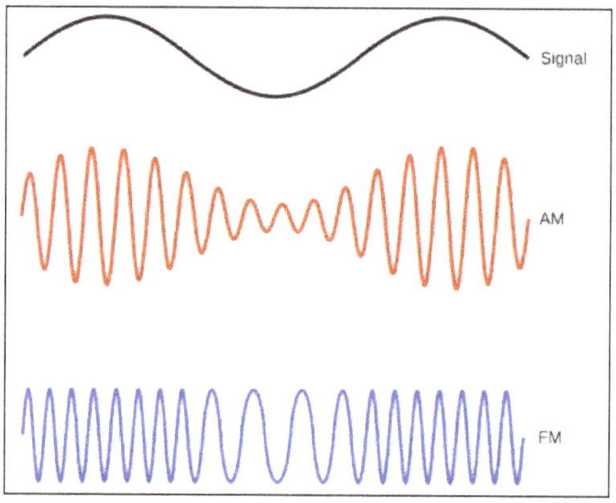

Original signal, AM and FM modulated signals.

In AM (Amplitude Modulation) the amplitude of the original signal (voice) is used to modulate the carrier and the transmitted signal has the information in the amplitude of the carrier wave. Whereas, in FM (Frequency Modulation) the frequency of the original signal (voice) is used to modulate the carrier and the transmitted signal has the information in the changing frequency of the carrier wave.

The first generation (1G) mobile cellular standards like AMPS (Advanced Mobile Phone System) and NMT (Nordic Mobile Telephone) were based on analog communication technologies. These were introduced in 1980s and were replaced by 2G standards which were based on digital cellular technologies by end 1990s or early 2000s.

AMPS phone.

Analog Transmission

Analog transmission is a transmission method of conveying information using a continuous signal which varies in amplitude, phase, or some other property in proportion to that information. It could be the transfer of an analog source signal, using an analog modulation method such as frequency modulation (FM) or amplitude modulation (AM), or no modulation at all.

Baseband data transmission using line codes, resulting in a pulse train, are always considered as digital transmission, although the source signal may be a digitized analog signal.

Methods of Transmission

Analog transmission can be conveyed in many different fashions:

- Optical fiber.
- Twisted pair or coaxial cable.
- Radio.
- Underwater acoustic communication.

There are two basic kinds of analog transmission, both based on how they modulate data to combine an input signal with a carrier signal. Usually, this carrier signal is a specific frequency, and data is transmitted through its variations. The two techniques are amplitude modulation (AM), which varies the amplitude of the carrier signal, and frequency modulation (FM), which modulates the frequency of the carrier.

Types of Analog Transmissions

Most analog transmissions fall into one of several categories. Telephony and voice communication was originally primarily analog in nature, as was most television and radio transmission. Early telecommunication devices utilized analog-to-digital conversion devices called modulator/demodulators, or modems, to convert analog signals to digital signals and back.

Benefits and Drawbacks

The analog transmission method is still very popular, in particular for shorter distances, due to significantly lower costs and complex multiplexing and timing equipment is unnecessary, and in small "short-haul" systems that simply do not need multiplexed digital transmission.

However, in situations where a signal often has high signal-to-noise ratio and cannot achieve source linearity, or in long distance, high output systems, analog is unattractive due to attenuation problems. Furthermore, as digital techniques continue to be refined, analog systems are increasingly becoming legacy equipment.

Recently, some nations, such as the Netherlands, have completely ceased analog transmissions (analogue switch-off) on certain media, such as television, for the purposes of the government saving money.

Amplitude Modulation

Amplitude modulation (AM) is a modulation technique used in electronic communication, most commonly for transmitting information via a radio carrier wave. In amplitude modulation, the amplitude (signal strength) of the carrier wave is varied in proportion to that of the message signal being transmitted. The message signal is, for example, a function of the sound to be reproduced by a loudspeaker, or the light intensity of pixels of a television screen. This technique contrasts with frequency modulation, in which the frequency of the carrier signal is varied, and phase modulation, in which its phase is varied.

AM was the earliest modulation method used for transmitting audio in radio broadcasting. It was developed during the first quarter of the 20th century beginning with Roberto Landell de Moura and Reginald Fessenden's radiotelephone experiments in 1900. It remains in use today in many forms of communication; for example, it is used in portable two-way radios, VHF aircraft radio, citizens band radio, and in computer modems in the form of QAM. AM is often used to refer to mediumwave AM radio broadcasting.

Forms of Amplitude Modulation

In electronics and telecommunications, modulation means varying some aspect of a continuous wave carrier signal with an information-bearing modulation waveform, such as an audio signal which represents sound, or a video signal which represents images. In this sense, the carrier wave, which has a much higher frequency than the message signal, carries the information. At the receiving station, the message signal is extracted from the modulated carrier by demodulation.

In amplitude modulation, the amplitude or strength of the carrier oscillations is varied. For example, in AM radio communication, a continuous wave radio-frequency signal (a sinusoidal carrier wave) has its amplitude modulated by an audio waveform before transmission. The audio waveform modifies the amplitude of the carrier wave and determines the envelope of the waveform. In the frequency domain, amplitude modulation produces a signal with power concentrated at the carrier frequency and two adjacent sidebands. Each sideband is equal in bandwidth to that of the modulating signal, and is a mirror image of the other. Standard AM is thus sometimes called "double-sideband amplitude modulation" (DSB-AM) to distinguish it from more sophisticated modulation methods also based on AM.

A disadvantage of all amplitude modulation techniques, not only standard AM, is that the receiver amplifies and detects noise and electromagnetic interference in equal proportion to the signal. Increasing the received signal-to-noise ratio, say, by a factor of 10 (a 10 decibel improvement), thus would require increasing the transmitter power by a factor of 10. This is in contrast to frequency modulation (FM) and digital radio where the effect of such noise following demodulation is strongly reduced so long as the received signal is well above the threshold for reception. For this reason AM broadcast is not favored for music and high fidelity broadcasting, but rather for voice communications and broadcasts (sports, news, talk radio etc).

AM is also inefficient in power usage; at least two-thirds of the power is concentrated in the carrier signal. The carrier signal contains none of the original information being transmitted (voice, video, data, etc). However its presence provides a simple means of demodulation using envelope detection, providing a frequency and phase reference to extract the modulation from the sidebands. In some modulation systems based on AM, a lower transmitter power is required through partial or total elimination of the carrier component; however receivers for these signals are more complex and costly. The receiver may regenerate a copy of the carrier frequency (usually as shifted to the intermediate frequency) from a greatly reduced "pilot" carrier (in reduced-carrier transmission or DSB-RC) to use in the demodulation process. Even with the carrier totally eliminated in double-sideband suppressed-carrier transmission, carrier regeneration is possible using a Costas phase-locked loop. This doesn't work however for single-sideband suppressed-carrier transmission (SSB-SC), leading to the characteristic "Donald Duck" sound from such receivers when slightly detuned. Single sideband is nevertheless used widely in amateur radio and other voice communications both due to its power efficiency and bandwidth efficiency (cutting the RF bandwidth in half compared to standard AM). On the other hand, in medium wave and short wave broadcasting, standard AM with the full carrier allows for reception using inexpensive receivers. The broadcaster absorbs the extra power cost to greatly increase potential audience.

An additional function provided by the carrier in standard AM, but which is lost in either single or double-sideband suppressed-carrier transmission, is that it provides an amplitude reference. In the receiver, the automatic gain control (AGC) responds to the carrier so that the reproduced audio level stays in a fixed proportion to the original modulation. On the other hand, with suppressed-carrier transmissions there is no transmitted power during pauses in the modulation, so the AGC must respond to peaks of the transmitted power during peaks in the modulation. This typically involves a so-called fast attack, slow decay circuit which holds the AGC level for a second or more following such peaks, in between syllables or short pauses in the program. This is very acceptable for communications radios, where compression of the audio aids intelligibility. However it is absolutely undesired for music or normal broadcast programming, where a faithful reproduction of the original program, including its varying modulation levels, is expected.

A trivial form of AM which can be used for transmitting binary data is on-off keying, the simplest form of amplitude-shift keying, in which ones and zeros are represented by the presence or absence of a carrier. On-off keying is likewise used by radio amateurs to transmit Morse code where it is known as continuous wave (CW) operation, even though the transmission is not strictly "continuous." A more complex form of AM, quadrature amplitude modulation is now more commonly used with digital data, while making more efficient use of the available bandwidth.

Simplified Analysis of Standard AM

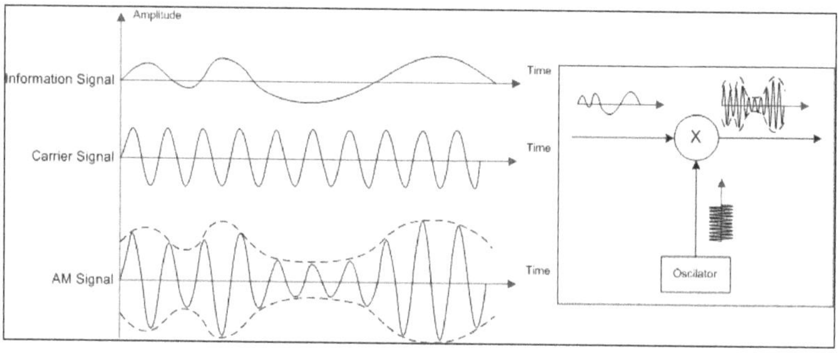

Illustration of amplitude modulation.

Consider a carrier wave (sine wave) of frequency fc and amplitude A given by:

$$c(t) = A\sin(2\pi f_c t).$$

Let $m(t)$ represent the modulation waveform. For this example we shall take the modulation to be simply a sine wave of a frequency f_m, a much lower frequency (such as an audio frequency) than f_c:

$$m(t) = M\cos(2\pi f_m t + \phi) = Am\cos(2\pi f_m t + \phi),$$

where m is the amplitude sensitivity, M is the amplitude of modulation. If $m < 1$, $(1 + m(t)/A)$ is always positive for undermodulation. If $m > 1$ then overmodulation occurs and reconstruction of message signal from the transmitted signal would lead in loss of original signal. Amplitude modulation results when the carrier $c(t)$ is multiplied by the positive quantity $(1 + m(t)/A)$:

$$y(t) = \left[1 + \frac{m(t)}{A}\right]c(t)$$
$$= \left[1 + m\cos(2\pi f_m t + \phi)\right]A\sin(2\pi f_c t)$$

In this simple case m is identical to the modulation index. With m = 0.5 the amplitude modulated signal y(t) thus corresponds to the top graph.

Using prosthaphaeresis identities, y(t) can be shown to be the sum of three sine waves:

$$y(t) = A\sin(2\pi f_c t) + \frac{1}{2}Am\left[\sin\left(2\pi[f_c + f_m]t + \phi\right) + \sin\left(2\pi[f_c - f_m]t - \phi\right)\right].$$

Therefore, the modulated signal has three components: the carrier wave c(t) which is unchanged, and two pure sine waves (known as sidebands) with frequencies slightly above and below the carrier frequency f_c.

Spectrum

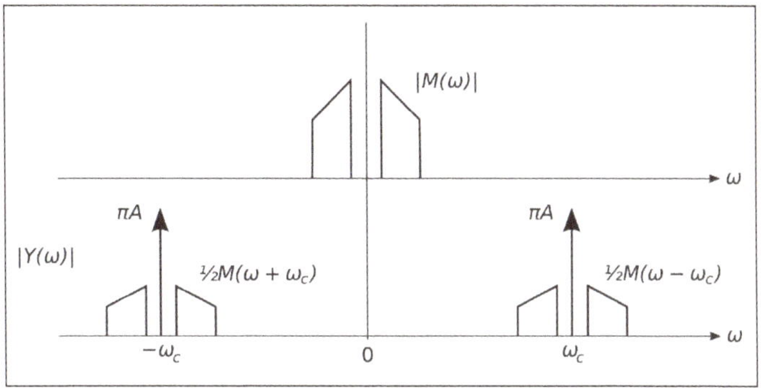

Double-sided spectra of baseband and AM signals.

A useful modulation signal m(t) is usually more complex than a single sine wave, as treated above. However, by the principle of Fourier decomposition, m(t) can be expressed as the sum of a set of sine waves of various frequencies, amplitudes, and phases. Carrying out the multiplication of 1 + m(t) with c(t) as above, the result consists of a sum of sine waves. Again, the carrier c(t) is present unchanged, but each frequency component of m at fi has two sidebands at frequencies fc + fi and fc - fi. The collection of the former frequencies above the carrier frequency is known as the upper sideband, and those below constitute the lower sideband. The modulation m(t) may be considered to consist of an equal mix of positive and negative frequency components. One can view the sidebands as that modulation m(t) having simply been shifted in frequency by f_c .

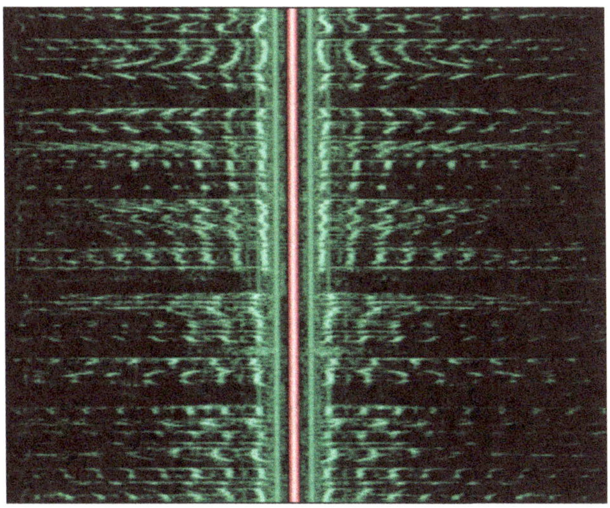

The spectrogram of an AM voice broadcast shows the two sidebands (green) on either side of the carrier (red) with time proceeding in the vertical direction.

The short-term spectrum of modulation, changing as it would for a human voice for instance, the frequency content (horizontal axis) may be plotted as a function of time (vertical axis). It can again be seen that as the modulation frequency content varies, an upper sideband is generated according to those frequencies shifted above the carrier frequency, and the same content mirror-imaged in the lower sideband below the carrier frequency. At all times, the carrier itself remains constant, and of greater power than the total sideband power.

Power and Spectrum Efficiency

The RF bandwidth of an AM transmission signal, since the upper and lower sidebands around the carrier frequency each have a bandwidth as wide as the highest modulating frequency. Although the bandwidth of an AM signal is narrower than one using frequency modulation (FM), it is twice as wide as single-sideband techniques; it thus may be viewed as spectrally inefficient. Within a frequency band, only half as many transmissions (or "channels") can thus be accommodated. For this reason analog television employs a variant of single-sideband (known as vestigial sideband, somewhat of a compromise in terms of bandwidth) in order to reduce the required channel spacing.

Another improvement over standard AM is obtained through reduction or suppression of the carrier component of the modulated spectrum. In figure this is the spike in between the sidebands; even with full (100%) sine wave modulation, the power in the carrier component is twice that in the sidebands, yet it carries no unique information. Thus there is a great advantage in efficiency in reducing or totally suppressing the carrier, either in conjunction with elimination of one sideband (single-sideband suppressed-carrier transmission) or with both sidebands remaining (double sideband suppressed carrier). While these suppressed carrier transmissions are efficient in terms of transmitter power, they require more sophisticated receivers employing synchronous detection and regeneration of the carrier frequency. For that reason, standard AM continues to be widely used, especially in broadcast transmission, to allow for the use of inexpensive receivers using envelope detection. Even (analog) television, with a (largely) suppressed lower sideband, includes sufficient carrier power for use of envelope detection. But for communications systems where both transmitters and receivers can be optimized, suppression of both one sideband and the carrier represent a net advantage and are frequently employed.

A technique used widely in broadcast AM transmitters is an application of the Hapburg carrier, first proposed in the 1930s but impractical with the technology then available. During periods of low modulation the carrier power would be reduced and would return to full power during periods of high modulation levels. This has the effect of reducing the overall power demand of the transmitter and is most effective on speech type programmes. Various trade names are used for its implementation by the transmitter manufacturers from the late 80's onwards.

Modulation Index

The AM modulation index is a measure based on the ratio of the modulation excursions of the RF signal to the level of the unmodulated carrier. It is thus defined as:

$$m = \frac{\text{peak value of } m(t)}{A} = \frac{M}{A}$$

where M, and A are the modulation amplitude and carrier amplitude, respectively; the modulation amplitude is the peak (positive or negative) change in the RF amplitude from its unmodulated value. Modulation index is normally expressed as a percentage, and may be displayed on a meter connected to an AM transmitter.

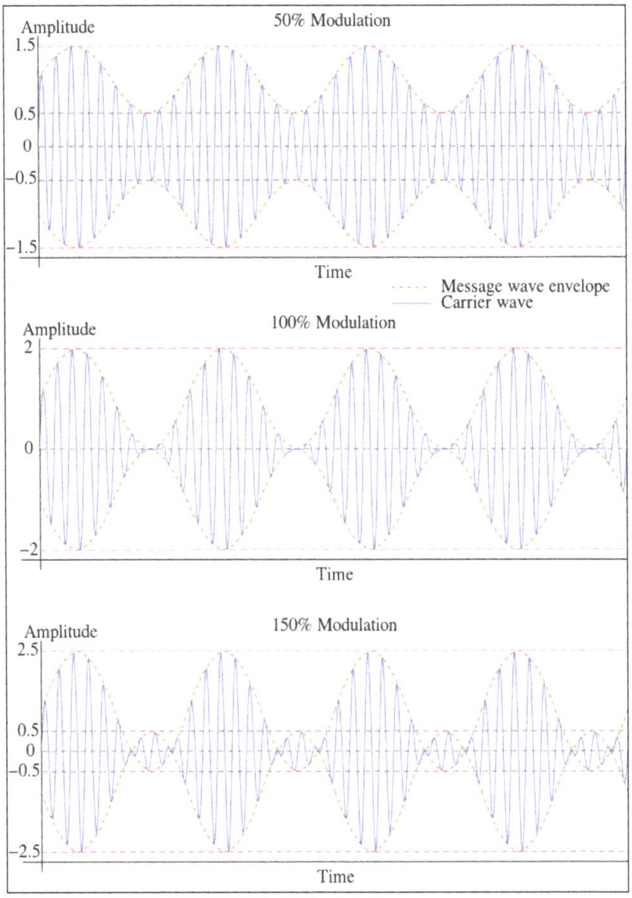

Modulation depth. In the diagram, the unmodulated carrier has an amplitude of 1.

So if $m = 0.5$, carrier amplitude varies by 50% above (and below) its unmodulated level, as is shown in the first waveform, below. For $m = 1.0$, it varies by 100% as shown in the illustration below it. With 100% modulation the wave amplitude sometimes reaches zero, and this represents full modulation using standard AM and is often a target (in order to obtain the highest possible signal-to-noise ratio) but mustn't be exceeded. Increasing the modulating signal beyond that point, known as overmodulation, causes a standard AM modulator to fail, as the negative excursions of the wave envelope cannot become less than zero, resulting in distortion ("clipping") of the received modulation. Transmitters typically incorporate a limiter circuit to avoid overmodulation, and/or a compressor circuit (especially for voice communications) in order to still approach 100% modulation for maximum intelligibility above the noise. Such circuits are sometimes referred to as a vogad.

However it is possible to talk about a modulation index exceeding 100%, without introducing distortion, in the case of double-sideband reduced-carrier transmission. In that case, negative excursions beyond zero entail a reversal of the carrier phase, as shown in the third waveform below. This cannot be produced using the efficient high-level (output stage) modulation techniques which are widely used especially in high power broadcast transmitters. Rather, a special modulator produces such a waveform at a low level followed by a linear amplifier. What's more, a standard AM receiver using an envelope detector is incapable of properly demodulating such a signal. Rather, synchronous detection is required. Thus double-sideband transmission is generally not referred to as "AM" even though it generates an identical RF waveform as standard AM as long as the modulation index is below 100%. Such systems more often attempt a radical reduction of the carrier level compared to the sidebands (where the useful information is present) to the point of double-sideband suppressed-carrier transmission where the carrier is (ideally) reduced to zero. In all such cases the term "modulation index" loses its value as it refers to the ratio of the modulation amplitude to a rather small (or zero) remaining carrier amplitude.

Modulation Methods

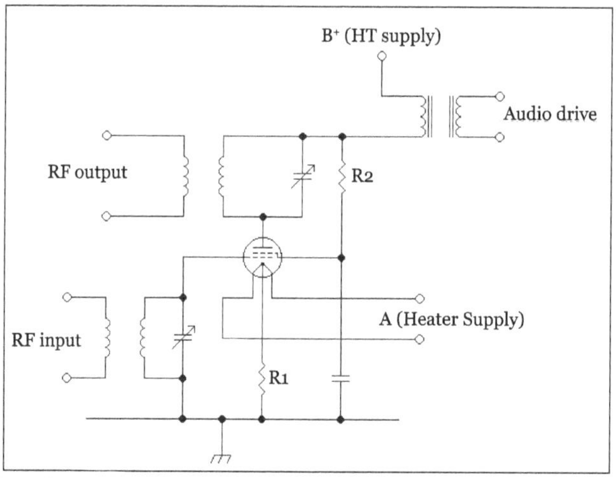

Anode (plate) modulation. A tetrode's plate and screen grid voltage is modulated via an audio transformer. The resistor R1 sets the grid bias; both the input and output are tuned circuits with inductive coupling.

Modulation circuit designs may be classified as low- or high-level (depending on whether they modulate in a low-power domain—followed by amplification for transmission—or in the high-power domain of the transmitted signal).

Low-level Generation

In modern radio systems, modulated signals are generated via digital signal processing (DSP). With DSP many types of AM are possible with software control (including DSB with carrier, SSB suppressed-carrier and independent sideband, or ISB). Calculated digital samples are converted to voltages with a digital-to-analog converter, typically at a frequency less than the desired RF-output frequency. The analog signal must then be shifted in frequency and linearly amplified to the desired frequency and power level (linear amplification must be used to prevent modulation distortion). This low-level method for AM is used in many Amateur Radio transceivers.

High-level Generation

High-power AM transmitters (such as those used for AM broadcasting) are based on high-efficiency class-D and class-E power amplifier stages, modulated by varying the supply voltage.

Older designs (for broadcast and amateur radio) also generate AM by controlling the gain of the transmitter's final amplifier (generally class-C, for efficiency). The following types are for vacuum tube transmitters (but similar options are available with transistors):

Plate Modulation

In plate modulation, the plate voltage of the RF amplifier is modulated with the audio signal. The audio power requirement is 50 percent of the RF-carrier power.

Heising (Constant-current) Modulation

RF amplifier plate voltage is fed through a choke (high-value inductor). The AM modulation tube plate is fed through the same inductor, so the modulator tube diverts current from the RF amplifier. The choke acts as a constant current source in the audio range. This system has a low power efficiency.

Control Grid Modulation

The operating bias and gain of the final RF amplifier can be controlled by varying the voltage of the control grid. This method requires little audio power, but care must be taken to reduce distortion.

Clamp Tube (Screen Grid) Modulation

The screen-grid bias may be controlled through a clamp tube, which reduces voltage according to the modulation signal. It is difficult to approach 100-percent modulation while maintaining low distortion with this system.

Doherty Modulation

One tube provides the power under carrier conditions and another operates only for positive modulation peaks. Overall efficiency is good, and distortion is low.

Outphasing Modulation

Two tubes are operated in parallel, but partially out of phase with each other. As they are differentially phase modulated their combined amplitude is greater or smaller. Efficiency is good and distortion low when properly adjusted.

Pulse-width Modulation (PWM) or Pulse-duration Modulation (PDM)

A highly efficient high voltage power supply is applied to the tube plate. The output voltage of this supply is varied at an audio rate to follow the program. This system was pioneered by Hilmer Swanson and has a number of variations, all of which achieve high efficiency and sound quality.

Demodulation Methods

The simplest form of AM demodulator consists of a diode which is configured to act as envelope detector. Another type of demodulator, the product detector, can provide better-quality demodulation with additional circuit complexity.

Quadrature Amplitude Modulation

Quadrature amplitude modulation (QAM) is the name of a family of digital modulation methods and a related family of analog modulation methods widely used in modern telecommunications to transmit information. It conveys two analog message signals, or two digital bit streams, by changing (modulating) the amplitudes of two carrier waves, using the amplitude-shift keying (ASK) digital modulation scheme or amplitude modulation (AM) analog modulation scheme. The two carrier waves of the same frequency are out of phase with each other by 90°, a condition known as orthogonality or quadrature. The transmitted signal is created by adding the two carrier waves together. At the receiver, the two waves can be coherently separated (demodulated) because of their orthogonality property. Another key property is that the modulations are low-frequency/low-bandwidth waveforms compared to the carrier frequency, which is known as the narrowband assumption.

Phase modulation (analog PM) and phase-shift keying (digital PSK) can be regarded as a special case of QAM, where the amplitude of the transmitted signal is a constant, but its phase varies. This can also be extended to frequency modulation (FM) and frequency-shift keying (FSK), for these can be regarded as a special case of phase modulation.

QAM is used extensively as a modulation scheme for digital telecommunication systems, such as in 802.11 Wi-Fi standards. Arbitrarily high spectral efficiencies can be achieved with QAM by setting a suitable constellation size, limited only by the noise level and linearity of the communications channel. QAM is being used in optical fiber systems as bit rates increase; QAM16 and QAM64 can be optically emulated with a 3-path interferometer.

In a QAM signal, one carrier lags the other by 90°, and its amplitude modulation is customarily referred to as the in-phase component, denoted by I(t). The other modulating function is the quadrature component, Q(t). So the composite waveform is mathematically modeled as:

$$s_s(t) \triangleq \sin(2\pi f_c t) I(t) + \underbrace{\sin\left(2\pi f_c t + \tfrac{\pi}{2}\right)}_{\cos(2\pi f_c t)} Q(t),$$

or,

$$s_c(t) \triangleq \cos(2\pi f_c t) I(t) + \underbrace{\cos\left(2\pi f_c t + \tfrac{\pi}{2}\right)}_{-\sin(2\pi f_c t)} Q(t),$$

where fc is the carrier frequency. At the receiver, a coherent demodulator multiplies the received signal separately with both a cosine and sine signal to produce the received estimates of I(t) and Q(t). For example:

$$r(t) \triangleq s_c(t)\cos(2\pi f_c t) = I(t)\cos(2\pi f_c t)\cos(2\pi f_c t) - Q(t)\sin(2\pi f_c t)\cos(2\pi f_c t)$$

Using standard trigonometric identities, we can write this as:

$$\begin{aligned} r(t) &= \tfrac{1}{2} I(t)\left[1 + \cos(4\pi f_c t)\right] - \tfrac{1}{2} Q(t)\sin(4\pi f_c t) \\ &= \tfrac{1}{2} I(t) + \tfrac{1}{2}[I(t)\cos(4\pi f_c t) - Q(t)\sin(4\pi f_c t)]. \end{aligned}$$

Low-pass filtering r(t) removes the high frequency terms (containing $4\pi f_c t$), leaving only the I(t) term. This filtered signal is unaffected by Q(t), showing that the in-phase component can be received independently of the quadrature component. Similarly, we can multiply sc(t) by a sine wave and then low-pass filter to extract Q(t).

The addition of two sinusoids is a linear operation that creates no new frequency components. So the bandwidth of the composite signal is comparable to the bandwidth of the DSB (Double-Sideband) components. Effectively, the spectral redundancy of DSB enables a doubling of the information capacity using this technique. This comes at the expense of demodulation complexity. In particular, a DSB signal has zero-crossings at a regular frequency, which makes it easy to recover the phase of the carrier sinusoid. It is said to be self-clocking. But the sender and receiver of a quadrature-modulated signal must share a clock or otherwise send a clock signal. If the clock phases drift apart, the demodulated I and Q signals bleed into each other, yielding crosstalk. In this context, the clock signal is called a "phase reference". Clock synchronization is typically achieved by transmitting a burst subcarrier or a pilot signal. The phase reference for NTSC, for example, is included within its colorburst signal.

Analog QAM is used in:

- NTSC and PAL analog color television systems, where the I- and Q-signals carry the components of chroma (colour) information. The QAM carrier phase is recovered from a special colorburst transmitted at the beginning of each scan line.
- C-QUAM ("Compatible QAM") is used in AM stereo radio to carry the stereo difference information.

Fourier Analysis of QAM

In the frequency domain, QAM has a similar spectral pattern to DSB-SC modulation. Applying Euler's formula to the sinusoids in $s_c(t) \triangleq \cos(2\pi f_c t)I(t) + \underbrace{\cos\left(2\pi f_c t + \tfrac{\pi}{2}\right)}_{-\sin(2\pi f_c t)} Q(t)$, the positive-frequency portion of sc (or analytic representation) is:

$$s_c(t)_+ = \tfrac{1}{2} e^{i2\pi f_c t}[I(t) + iQ(t)] \overset{\mathcal{F}}{\Rightarrow} \tfrac{1}{2}\left[\widehat{I}(f - f_c) + e^{i\pi/2}\widehat{Q}(f - f_c)\right],$$

where $\mathcal{F}$ denotes the Fourier transform, and $\hat{\text{I}}$ and $\widehat{\text{Q}}$ are the transforms of I(t) and Q(t). This result represents the sum of two DSB-SC signals with the same center frequency. The factor of i ($= e^{i\pi/2}$)represents the 90° phase shift that enables their individual demodulations.

Digital QAM

As in many digital modulation schemes, the constellation diagram is useful for QAM. In QAM, the constellation points are usually arranged in a square grid with equal vertical and horizontal spacing, although other configurations are possible (e.g. Cross-QAM). Since in digital telecommunications the data is usually binary, the number of points in the grid is usually a power of 2 (2, 4, 8, ...). Since QAM is usually square, some of these are rare—the most common forms are 16-QAM, 64-QAM and 256-QAM. By moving to a higher-order constellation, it is possible to transmit more bits per symbol. However, if the mean energy of the constellation is to remain the same (by way of making a fair comparison), the points must be closer together and are thus more susceptible to noise and other corruption; this results in a higher bit error rate and so higher-order QAM can deliver more data less reliably than lower-order QAM, for constant mean constellation energy. Using higher-order QAM without increasing the bit error rate requires a higher signal-to-noise ratio (SNR) by increasing signal energy, reducing noise, or both.

If data-rates beyond those offered by 8-PSK are required, it is more usual to move to QAM since it achieves a greater distance between adjacent points in the I-Q plane by distributing the points more evenly. The complicating factor is that the points are no longer all the same amplitude and so the demodulator must now correctly detect both phase and amplitude, rather than just phase.

64-QAM and 256-QAM are often used in digital cable television and cable modem applications. In the United States, 64-QAM and 256-QAM are the mandated modulation schemes for digital cable (see QAM tuner) as standardised by the SCTE in the standard ANSI/SCTE 07 2013. Note that many marketing people will refer to these as QAM-64 and QAM-256.In the UK, 64-QAM is used for digital terrestrial television (Freeview) whilst 256-QAM is used for Freeview-HD.

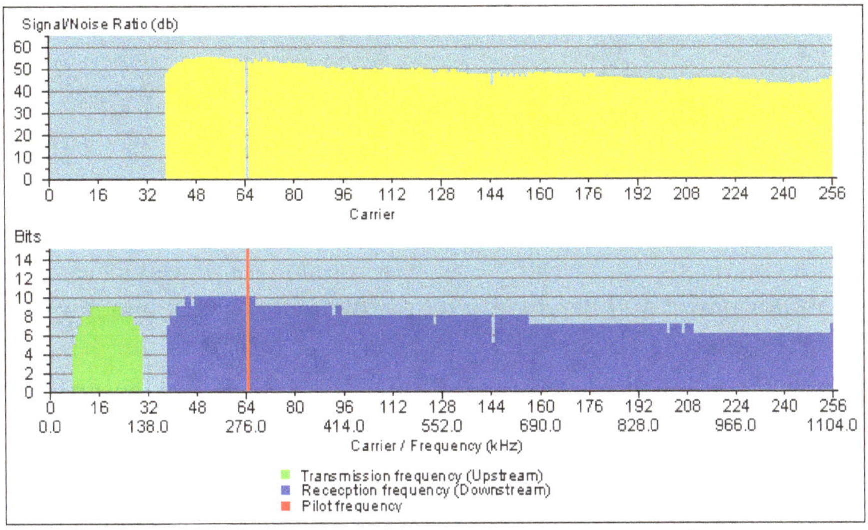

Bit-loading (bits per QAM constellation) on an ADSL line.

Communication systems designed to achieve very high levels of spectral efficiency usually employ very dense QAM constellations. For example, current Homeplug AV2 500-Mbit/s powerline Ethernet devices use 1024-QAM and 4096-QAM, as well as future devices using ITU-T G.hn standard for networking over existing home wiring (coaxial cable, phone lines and power lines); 4096-QAM provides 12 bits/symbol. Another example is ADSL technology for copper twisted pairs, whose constellation size goes up to 32768-QAM (in ADSL terminology this is referred to as bit-loading, or bit per tone, 32768-QAM being equivalent to 15 bits per tone).

Ultra-high capacity Microwave Backhaul Systems also use 1024-QAM. With 1024-QAM, adaptive coding and modulation (ACM) and XPIC, vendors can obtain gigabit capacity in a single 56 MHz channel.

Interference and Noise

In moving to a higher order QAM constellation (higher data rate and mode) in hostile

RF/microwave QAM application environments, such as in broadcasting or telecommunications, multipath interference typically increases. There is a spreading of the spots in the constellation, decreasing the separation between adjacent states, making it difficult for the receiver to decode the signal appropriately. In other words, there is reduced noise immunity. There are several test parameter measurements which help determine an optimal QAM mode for a specific operating environment. The following three are most significant:

- Carrier/interference ratio.
- Carrier-to-noise ratio.
- Threshold-to-noise ratio.

Single-sideband Modulation

In radio communications, single-sideband modulation (SSB) or single-sideband suppressed-carrier modulation (SSB-SC) is a type of modulation used to transmit information, such as an audio signal, by radio waves. A refinement of amplitude modulation, it uses transmitter power and bandwidth more efficiently. Amplitude modulation produces an output signal the bandwidth of which is twice the maximum frequency of the original baseband signal. Single-sideband modulation avoids this bandwidth increase, and the power wasted on a carrier, at the cost of increased device complexity and more difficult tuning at the receiver.

Radio transmitters work by mixing a radio frequency (RF) signal of a specific frequency, the carrier wave, with the audio signal to be broadcast. In AM transmitters this mixing usually takes place in the final RF amplifier (high level modulation). It is less common and much less efficient to do the mixing at low power and then amplify it in a linear amplifier. Either method produces a set of frequencies with a strong signal at the carrier frequency and with weaker signals at frequencies extending above and below the carrier frequency by the maximum frequency of the input signal. Thus the resulting signal has a spectrum whose bandwidth is twice the maximum frequency of the original input audio signal.

SSB takes advantage of the fact that the entire original signal is encoded in each of these "sidebands". It is not necessary to transmit both sidebands plus the carrier, as a suitable receiver can extract the entire original signal from either the upper or lower sideband. There are several methods for eliminating the carrier and one sideband from the transmitted signal. Producing this single sideband signal is too complicated to be done in the final amplifier stage as with AM. SSB Modulation must be done at a low level and amplified in a linear amplifier where lower efficiency partially offsets the power advantage gained by eliminating the carrier and one sideband. Nevertheless, SSB transmissions use the available amplifier energy considerably more efficiently, providing longer-range transmission for the same power output. In addition, the occupied spectrum is less than half that of a full carrier AM signal.

SSB reception requires frequency stability and selectivity well beyond that of inexpensive AM receivers which is why broadcasters have seldom used it. In point to point communications where expensive receivers are in common use already they can successfully be adjusted to receive whichever sideband is being transmitted.

Mathematical Formulation

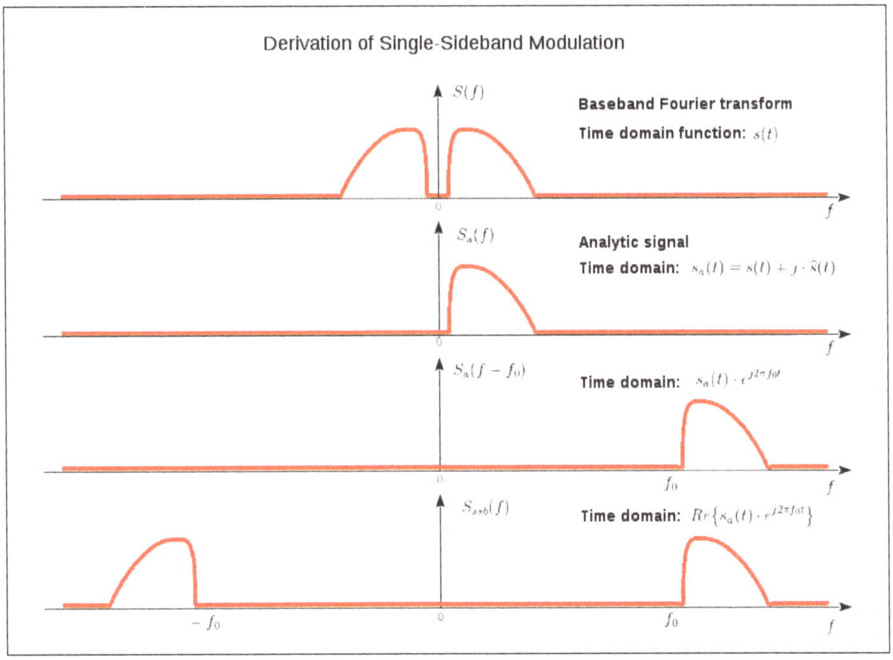

Frequency-domain depiction of the mathematical steps that convert a baseband function into a single-sideband radio signal.

Single-sideband has the mathematical form of quadrature amplitude modulation (QAM) in the special case where one of the baseband waveforms is derived from the other, instead of being independent messages:

$$s_{\mathrm{ssb}}(t) = s(t)\cdot\cos(2\pi f_0 t) - \hat{s}(t)\cdot\sin(2\pi f_0 t),$$

where $s(t)$ is the message (real-valued), $\hat{s}(t)$ is its Hilbert transform, and f_0 is the radio carrier frequency.

To understand this formula, we may express $s(t)$ as the real part of a complex-valued function, with no loss of information:

$$s(t) = \mathrm{Re}\{s_{\mathrm{a}}(t)\} = \mathrm{Re}\{s(t) + j\cdot\hat{s}(t)\},$$

where j represents the imaginary unit. $s_{\mathrm{a}}(t)$ is the analytic representation of $s(t)$, which means that it comprises only the positive-frequency components of $s(t)$:

$$\frac{1}{2}S_{\mathrm{a}}(f)=\begin{cases}S(f), & \text{for } f>0,\\ 0, & \text{for } f<0,\end{cases}$$

where $S_{\mathrm{a}}(f)$ and $S(f)$ are the respective Fourier transforms of $s_{\mathrm{a}}(t)$ and $s(t)$. Therefore, the frequency-translated function $S_{\mathrm{a}}(f-f_0)$ contains only one side of $S(f)$. Since it also has only positive-frequency components, its inverse Fourier transform is the analytic representation of $s_{\mathrm{ssb}}(t)$:

$$s_{\mathrm{ssb}}(t)+j\cdot\hat{s}_{\mathrm{ssb}}(t)=\mathcal{F}^{-1}\{S_{\mathrm{a}}(f-f_0)\}=s_{\mathrm{a}}(t)\cdot e^{j2\pi f_0 t},$$

and again the real part of this expression causes no loss of information. With Euler's formula to expand $e^{j2\pi f_0 t}$, we obtain $s_{\mathrm{ssb}}(t)=s(t)\cdot\cos(2\pi f_0 t)-\hat{s}(t)\cdot\sin(2\pi f_0 t)$,

$$\begin{aligned}s_{\mathrm{ssb}}(t)&=\mathrm{Re}\left\{s_{\mathrm{a}}(t)\cdot e^{j2\pi f_0 t}\right\}\\&=\mathrm{Re}\left\{\left[s(t)+j\cdot\hat{s}(t)\right]\cdot\left[\cos(2\pi f_0 t)+j\cdot\sin(2\pi f_0 t)\right]\right\}\\&=s(t)\cdot\cos(2\pi f_0 t)-\hat{s}(t)\cdot\sin(2\pi f_0 t).\end{aligned}$$

Coherent demodulation of $s_{\mathrm{ssb}}(t)$ to recover $s(t)$ is the same as AM: multiply by $\cos(2\pi f_0 t)$, and lowpass to remove the "double-frequency" components around frequency $2f_0$. If the demodulating carrier is not in the correct phase (cosine phase here), then the demodulated signal will be some linear combination of $s(t)$ and $\hat{s}(t)$, which is usually acceptable in voice communications (if the demodulation carrier frequency is not quite right, the phase will be drifting cyclically, which again is usually acceptable in voice communications if the frequency error is small enough, and amateur radio operators are sometimes tolerant of even larger frequency errors that cause unnatural-sounding pitch shifting effects).

Lower Sideband

$s(t)$ can also be recovered as the real part of the complex-conjugate $s_{\mathrm{a}}^{*}(t)$, which represents the negative frequency portion of $S(f)$. When f_0, is large enough that $S(f-f_0)$ has no negative frequencies $s_{\mathrm{a}}^{*}(t)\cdot e^{j2\pi f_0 t}$, the product is another analytic signal, whose real part is the actual lower-sideband transmission:

$$\begin{aligned}&s_{\mathrm{a}}^{*}(t)\cdot e^{j2\pi f_0 t}=s_{\mathrm{lsb}}(t)+j\cdot\hat{s}_{\mathrm{lsb}}(t)\\&\Rightarrow s_{\mathrm{lsb}}(t)=\mathrm{Re}\left\{s_{\mathrm{a}}^{*}(t)\cdot e^{j2\pi f_0 t}\right\}\\&\qquad\qquad=s(t)\cdot\cos(2\pi f_0 t)+\hat{s}(t)\cdot\sin(2\pi f_0 t)\end{aligned}$$

The sum of the two sideband signals is:

$$s_{\mathrm{usb}}(t)+s_{\mathrm{lsb}}(t)=2s(t)\cdot\cos\left(2\pi f_0 t\right),$$

which is the classic model of suppressed-carrier double sideband AM.

Practical Implementations

A Collins KWM-1, an early Amateur Radio transceiver that featured SSB voice capability.

Bandpass Filtering

One method of producing an SSB signal is to remove one of the sidebands via filtering, leaving only either the upper sideband (USB), the sideband with the higher frequency, or less commonly the lower sideband (LSB), the sideband with the lower frequency. Most often, the carrier is reduced or removed entirely (suppressed), being referred to in full as single sideband suppressed carrier (SSBSC). Assuming both sidebands are symmetric, which is the case for a normal AM signal, no information is lost in the process. Since the final RF amplification is now concentrated in a single sideband, the effective power output is greater than in normal AM (the carrier and redundant sideband account for well over half of the power output of an AM transmitter). Though SSB uses substantially less bandwidth and power, it cannot be demodulated by a simple envelope detector like standard AM.

Hartley Modulator

An alternate method of generation known as a Hartley modulator, named after R. V. L. Hartley, uses phasing to suppress the unwanted sideband. To generate an SSB signal with this method, two versions of the original signal are generated, mutually 90° out of phase for any single frequency within the operating bandwidth. Each one of these signals then modulates carrier waves (of one frequency) that are also 90° out of phase with each other. By either adding or subtracting the resulting signals, a lower or upper sideband signal results. A benefit of this approach is to allow an analytical expression for SSB signals, which can be used to understand effects such as synchronous detection of SSB.

Shifting the baseband signal 90° out of phase cannot be done simply by delaying it, as it contains a large range of frequencies. In analog circuits, a wideband 90-degree phase-difference network is used. The method was popular in the days of vacuum tube radios, but later gained a bad reputation due to poorly adjusted commercial implementations. Modulation using this method is again gaining popularity in the homebrew and DSP fields. This method, utilizing the Hilbert transform to phase shift the baseband audio, can be done at low cost with digital circuitry.

Weaver Modulator

Another variation, the Weaver modulator, uses only lowpass filters and quadrature mixers, and is a favored method in digital implementations.

In Weaver's method, the band of interest is first translated to be centered at zero, conceptually by modulating a complex exponential $\exp(j\omega t)$ with frequency in the middle of the voiceband, but implemented by a quadrature pair of sine and cosine modulators at that frequency (e.g. 2 kHz). This complex signal or pair of real signals is then lowpass filtered to remove the undesired sideband that is not centered at zero. Then, the single-sideband complex signal centered at zero is upconverted to a real signal, by another pair of quadrature mixers, to the desired center frequency.

Full, Reduced and Suppressed-carrier SSB

Conventional amplitude-modulated signals can be considered wasteful of power and bandwidth because they contain a carrier signal and two identical sidebands. Therefore, SSB transmitters are generally designed to minimize the amplitude of the carrier signal. When the carrier is removed from the transmitted signal, it is called suppressed-carrier SSB.

However, in order for a receiver to reproduce the transmitted audio without distortion, it must be tuned to exactly the same frequency as the transmitter. Since this is difficult to achieve in practice, SSB transmissions can sound unnatural, and if the error in frequency is great enough, it can cause poor intelligibility. In order to correct this, a small amount of the original carrier signal can be transmitted so that receivers with the necessary circuitry to synchronize with the transmitted carrier can correctly demodulate the audio. This mode of transmission is called reduced-carrier single-sideband.

In other cases, it may be desirable to maintain some degree of compatibility with simple AM receivers, while still reducing the signal's bandwidth. This can be accomplished by transmitting single-sideband with a normal or slightly reduced carrier. This mode is called compatible (or full-carrier) SSB or amplitude modulation equivalent (AME). In typical AME systems, harmonic distortion can reach 25%, and intermodulation distortion can be much higher than normal, but minimizing distortion in receivers with envelope detectors is generally considered less important than allowing them to produce intelligible audio.

A second, and perhaps more correct, definition of "compatible single sideband" (CSSB) refers to a form of amplitude and phase modulation in which the carrier is transmitted along with a series of sidebands that are predominantly above or below the carrier term. Since phase modulation is present in the generation of the signal, energy is removed from the carrier term and redistributed into the sideband structure similar to that which occurs in analog frequency modulation. The signals feeding the phase modulator and the envelope modulator are further phase-shifted by 90° with respect to each other. This places the information terms in quadrature with each other; the Hilbert transform of information to be transmitted is utilized to cause constructive addition of one sideband and cancellation of the opposite primary sideband. Since phase modulation is employed, higher-order terms are also generated. Several methods have been employed to reduce the impact (amplitude) of most of these higher-order terms. In one system, the phase-modulated term is actually the log of the value of the carrier level plus the phase-shifted audio/information term. This produces an ideal CSSB signal, where at low modulation levels only a first-order term on one side of the carrier is predominant. As the modulation level is increased, the carrier level is reduced while a second-order term increases substantially in amplitude. At the point of 100% envelope modulation, 6 dB of power is removed from the carrier term, and the second-order term is identical in amplitude to carrier term.

The first-order sideband has increased in level until it is now at the same level as the formerly unmodulated carrier. At the point of 100% modulation, the spectrum appears identical to a normal double-sideband AM transmission, with the center term (now the primary audio term) at a 0 dB reference level, and both terms on either side of the primary sideband at −6 dB. The difference is that what appears to be the carrier has shifted by the audio-frequency term towards the "sideband in use". At levels below 100% modulation, the sideband structure appears quite asymmetric. When voice is conveyed by a CSSB source of this type, low-frequency components are dominant, while higher-frequency terms are lower by as much as 20 dB at 3 kHz. The result is that the signal occupies approximately 1/2 the normal bandwidth of a full-carrier, DSB signal. There is one catch: the audio term utilized to phase-modulate the carrier is generated based on a log function that is biased by the carrier level. At negative 100% modulation, the term is driven to zero (0), and the modulator becomes undefined. Strict modulation control must be employed to maintain stability of the system and avoid splatter. This system is of Russian origin and was described in the late 1950s. It is uncertain whether it was ever deployed.

Demodulation

The front end of an SSB receiver is similar to that of an AM or FM receiver, consisting of a superheterodyne RF front end that produces a frequency-shifted version of the radio frequency (RF) signal within a standard intermediate frequency (IF) band.

To recover the original signal from the IF SSB signal, the single sideband must be frequency-shifted down to its original range of baseband frequencies, by using a product detector which mixes it with the output of a beat frequency oscillator (BFO). In other words, it is just another stage of heterodyning. For this to work, the BFO frequency must be exactly adjusted. If the BFO frequency is off, the output signal will be frequency-shifted (up or down), making speech sound strange and "Donald Duck"-like, or unintelligible.

For audio communications, there is a common agreement about the BFO oscillator shift of 1.7 kHz. A voice signal is sensitive to about 50 Hz shift, with up to 100 Hz still bearable. Some receivers use a carrier recovery system, which attempts to automatically lock on to the exact IF frequency. The carrier recovery doesn't solve the frequency shift. It gives better S/N ratio on the detector output.

As an example, consider an IF SSB signal centered at frequency F_{if} = 45000 Hz. The baseband frequency it needs to be shifted to is F_b = 2000 Hz. The BFO output waveform is $\cos(2\pi \cdot F_{\text{bfo}} \cdot t)$. When the signal is multiplied by (aka heterodyned with) the BFO waveform, it shifts the signal to $(F_{\text{if}} + F_{\text{bfo}})$, and to $|F_{\text{if}} - F_{\text{bfo}}|$, which is known as the *beat frequency* or *image frequency*. The objective is to choose an F_{BFO} that results in $|F_{\text{if}} - F_{\text{bfo}}| = F_b = 2000$ Hz. (The unwanted components at $(F_{\text{if}} + F_{\text{bfo}})$ can be removed by a lowpass filter; for which an output transducer or the human ear may serve).

There are two choices for F_{bfo}: 43000 Hz and 47000 Hz, called low-side and high-side injection. With high-side injection, the spectral components that were distributed around 45000 Hz will be distributed around 2000 Hz in the reverse order, also known as an inverted spectrum. That is in fact desirable when the IF spectrum is also inverted, because the BFO inversion restores the proper relationships. One reason for that is when the IF spectrum is the output of an inverting stage in the receiver. Another reason is when the SSB signal is actually a lower sideband, instead of an upper sideband. But if both reasons are true, then the IF spectrum is not inverted, and the non-inverting BFO (43000 Hz) should be used.

If F_{bfo} is off by a small amount, then the beat frequency is not exactly F_b, which can lead to the speech distortion.

SSB as a Speech-scrambling Technique

SSB techniques can also be adapted to frequency-shift and frequency-invert baseband waveforms (voice inversion). This voice scrambling method was made by running the audio of one side band modulated audio sample though its opposite (e.g. running an LSB modulated audio sample through a radio running USB modulation). These effects were used, in conjunction with other filtering techniques, during World War II as a simple method for speech encryption. Radiotelephone conversations between the US and Britain were intercepted and "decrypted" by the Germans; they included some early conversations between Franklin D. Roosevelt and Churchill. In fact, the signals could be understood directly by trained operators. Largely to allow secure communications between Roosevelt and Churchill, the SIGSALY system of digital encryption was devised.

Today, such simple inversion-based speech encryption techniques are easily decrypted using simple techniques and are no longer regarded as secure.

Vestigial Sideband (VSB)

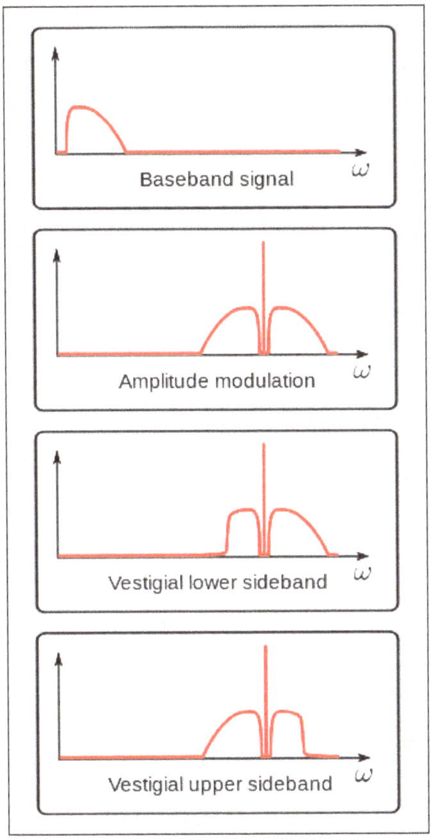

VSB Modulation.

Limitation of single-sideband modulation being used for voice signals and not available for video/TV signals leads to the usage of vestigial sideband. A vestigial sideband (in radio communication) is a sideband that has been only partly cut off or suppressed. Television broadcasts (in analog video formats) use this method if the video is transmitted in AM, due to the large bandwidth used. It may also be used in digital transmission, such as the ATSC standardized 8VSB.

The broadcast or transport channel for TV in countries that use NTSC or ATSC has a bandwidth of 6 MHz. To conserve bandwidth, SSB would be desirable, but the video signal has significant low-frequency content (average brightness) and has rectangular synchronising pulses. The engineering compromise is vestigial-sideband transmission. In vestigial sideband, the full upper sideband of bandwidth W2 = 4.75 MHz is transmitted, but only W1 = 1.25 MHz of the lower sideband is transmitted, along with a carrier. This effectively makes the system AM at low modulation frequencies and SSB at high modulation frequencies. The absence of the lower sideband components at high frequencies must be compensated for, and this is done in the IF amplifier.

Frequencies for LSB and USB in Amateur Radio Voice Communication

When single-sideband is used in amateur radio voice communications, it is common practice that for frequencies below 10 MHz, lower sideband (LSB) is used and for frequencies of 10 MHz and above, upper sideband (USB) is used. For example, on the 40 m band, voice communications often take place around 7.100 MHz using LSB mode. On the 20 m band at 14.200 MHz, USB mode would be used.

An exception to this rule applies to the five discrete amateur channels on the 60-meter band (near 5.3 MHz) where FCC rules specifically require USB.

Extended Single Sideband (eSSB)

Extended single sideband is any J3E (SSB-SC) mode that exceeds the audio bandwidth of standard or traditional 2.9 kHz SSB J3E modes (ITU 2K90J3E) to support higher-quality sound.

Extended SSB modes	Bandwidth	Frequency response	ITU Designator
eSSB (Narrow-1a)	3 kHz	100 Hz ~ 3.10 kHz	3K00J3E
eSSB (Narrow-1b)	3 kHz	50 Hz ~ 3.05 kHz	3K00J3E
eSSB (Narrow-2)	3.5 kHz	50 Hz ~ 3.55 kHz	3K50J3E
eSSB (Medium-1)	4 kHz	50 Hz ~ 4.05 kHz	4K00J3E
eSSB (Medium-2)	4.5 kHz	50 Hz ~ 4.55 kHz	4K50J3E
eSSB (Wide-1)	5 kHz	50 Hz ~ 5.05 kHz	5K00J3E
eSSB (Wide-2)	6 kHz	50 Hz ~ 6.05 kHz	6K00J3E

Amplitude-companded Single-sideband Modulation (ACSSB)

Amplitude-companded single sideband (ACSSB) is a narrowband modulation method using a single sideband with a pilot tone, allowing an expander in the receiver to restore the amplitude that was severely compressed by the transmitter. It offers improved effective range over standard SSB modulation while simultaneously retaining backwards compatibility with standard SSB radios. ACSSB also offers reduced bandwidth and improved range for a given power level compared with narrow band FM modulation.

Controlled-envelope Single-sideband Modulation (CESSB)

The generation of standard SSB modulation results in large envelope overshoots well above the average envelope level for a sinusoidal tone (even when the audio signal is peak-limited). The standard SSB envelope peaks are due to truncation of the spectrum and nonlinear phase distortion from the approximation errors of the practical implementation of the required Hilbert transform. It was recently shown that suitable overshoot compensation (so-called controlled-envelope single-sideband modulation or CESSB) achieves about 3.8 dB of peak reduction for speech transmission. This results in an effective average power increase of about 140%. Although the generation of the CESSB signal can be integrated into the SSB modulator, it is feasible to separate the generation of the CESSB signal (e.g. in form of an external speech preprocessor) from a standard SSB radio. This requires that the standard SSB radio's modulator be linear-phase and have a sufficient bandwidth to pass the CESSB signal. If a standard SSB modulator meets these requirements, then the envelope control by the CESSB process is preserved.

ITU designations

In 1982, the International Telecommunication Union (ITU) designated the types of amplitude modulation:

Designation	Description
A3E	Double-sideband full-carrier – the basic amplitude-modulation scheme
R3E	Single-sideband reduced-carrier
H3E	Single-sideband full-carrier
J3E	Single-sideband suppressed-carrier
B8E	Independent-sideband emission
C3F	Vestigial-sideband
Lincompex	Linked compressor and expander

Space Modulation

Space modulation is a radio amplitude modulation technique used in instrument landing systems (ILS) that incorporates the use of multiple antennas fed with various radio frequency powers and phases to create different depths of modulation within various volumes of three-dimensional airspace. This modulation method differs from internal modulation methods inside most other radio transmitters in that the phases and powers of the two individual signals mix within airspace, rather than in a modulator.

An aircraft with an on-board ILS receiver within the capture area of an ILS, (glideslope and localizer range), will detect varying depths of modulation according to the aircraft's position within that airspace, providing accurate positional information about the progress to the threshold.

Method used to Determine Aircraft Position

The ILS uses two radio frequencies, one for each ground station (about 110 MHz for LOC and 330 MHz for the GS), to transmit two amplitude-modulated signals (90 Hz and 150 Hz), along the glidepath (GS) and the course (LOC) trajectories into airspace. It is this signal that is projected up from the runway which an aircraft employing an instrument approach uses to land.

The modulation depth of each 90 Hz and 150 Hz signal changes according to the deviation of the aircraft from the correct position for the aircraft to touchdown on the threshold. The difference between the two signal modulation depths is zero when the aircraft is on the correct course and glidepath on approach to the runway—i.e. No difference (zero DDM), produces no deviation from the middle indication of the instrument's needle within the cockpit of the aircraft.

Angle Modulation

The other type of modulation in continuous-wave modulation is the Angle Modulation. Angle Modulation is the process in which the frequency or the phase of the carrier varies according to the message signal. This is further divided into frequency and phase modulation.

- Frequency Modulation is the process of varying the frequency of the carrier signal linearly with the message signal.
- Phase Modulation is the process of varying the phase of the carrier signal linearly with the message signal.

Frequency Modulation (FM)

In telecommunications and signal processing, frequency modulation (FM) is the encoding of information in a carrier wave by varying the instantaneous frequency of the wave.

In analog frequency modulation, such as FM radio broadcasting of an audio signal representing voice or music, the instantaneous frequency deviation, the difference between the frequency of the carrier and its center frequency, is proportional to the modulating signal.

Digital data can be encoded and transmitted via FM by shifting the carrier's frequency among a predefined set of frequencies representing digits – for example one frequency can represent a binary 1 and a second can represent binary 0. This modulation technique is known as frequency-shift keying (FSK). FSK is widely used in modems such as fax modems, and can also be used to send Morse code. Radioteletype also uses FSK.

Frequency modulation is widely used for FM radio broadcasting. It is also used in telemetry, radar, seismic prospecting, and monitoring newborns for seizures via EEG, two-way radio systems, music synthesis, magnetic tape-recording systems and some video-transmission systems. In radio transmission, an advantage of frequency modulation is that it has a larger signal-to-noise ratio and therefore rejects radio frequency interference better than an equal power amplitude modulation (AM) signal. For this reason, most music is broadcast over FM radio.

Frequency modulation and phase modulation are the two complementary principal methods of angle modulation; phase modulation is often used as an intermediate step to achieve frequency modulation. These methods contrast with amplitude modulation, in which the amplitude of the carrier wave varies, while the frequency and phase remain constant.

If the information to be transmitted (i.e., the baseband signal) is $x_m(t)$ and the sinusoidal carrier is $x_c(t) = A_c \cos(2\pi f_c t)$, where f_c is the carrier's base frequency, and A_c is the carrier's amplitude, the modulator combines the carrier with the baseband data signal to get the transmitted signal:

$$\begin{aligned} y(t) &= A_c \cos\left(2\pi \int_0^t f(\tau) d\tau\right) \\ &= A_c \cos\left(2\pi \int_0^t [f_c + f_\Delta x_m(\tau)] d\tau\right) \\ &= A_c \cos\left(2\pi f_c t + 2\pi f_\Delta \int_0^t x_m(\tau) d\tau\right) \end{aligned}$$

where $f_\Delta = K_f A_m, K_f$ being the sensitivity of the frequency modulator and A_m being the amplitude of the modulating signal or baseband signal.

In this equation, $f(\tau)$ is the instantaneous frequency of the oscillator and f_Δ is the frequency deviation, which represents the maximum shift away from fc in one direction, assuming xm(t) is limited to the range ±1.

While most of the energy of the signal is contained within fc ± fΔ, it can be shown by Fourier analysis that a wider range of frequencies is required to precisely represent an FM signal. The frequency spectrum of an actual FM signal has components extending infinitely, although their amplitude decreases and higher-order components are often neglected in practical design problems.

Sinusoidal Baseband Signal

Mathematically, a baseband modulating signal may be approximated by a sinusoidal continuous wave signal with a frequency fm. This method is also named as single-tone modulation. The integral of such a signal is:

$$\int_0^t x_m(\tau)d\tau = A_m \frac{\sin(2\pi f_m t)}{2\pi f_m}$$

In this case, the expression for y(t) above simplifies to:

$$y(t) = A_c \cos\left(2\pi f_c t + \frac{A_m f_\Delta}{f_m}\sin(2\pi f_m t)\right)$$

where the amplitude A_m of the modulating sinusoid is represented by the peak deviation f_Δ.

The harmonic distribution of a sine wave carrier modulated by such a sinusoidal signal can be represented with Bessel functions; this provides the basis for a mathematical understanding of frequency modulation in the frequency domain.

Modulation Index

As in other modulation systems, the modulation index indicates by how much the modulated variable varies around its unmodulated level. It relates to variations in the carrier frequency:

$$h = \frac{\Delta f}{f_m} = \frac{f_\Delta |x_m(t)|}{f_m}$$

where f_m is the highest frequency component present in the modulating signal $x_m(t)$, and Δf is the peak frequency-deviation—i.e. the maximum deviation of the instantaneous frequency from the carrier frequency. For a sine wave modulation, the modulation index is seen to be the ratio of the peak frequency deviation of the carrier wave to the frequency of the modulating sine wave.

If $h \ll 1$, the modulation is called narrowband FM (NFM), and its bandwidth is approximately $2f_m$. Sometimes modulation index $h < 0.3$ is considered as NFM, otherwise wideband FM (WFM or FM).

For digital modulation systems, for example binary frequency shift keying (BFSK), where a binary signal modulates the carrier, the modulation index is given by:

$$h = \frac{\Delta f}{f_m} = \frac{\Delta f}{\frac{1}{2T_s}} = 2\Delta f T_s$$

where T_s is the symbol period, and $f_m = \frac{1}{2T_s}$ is used as the highest frequency of the modulating binary waveform by convention, even though it would be more accurate to say it is the highest fundamental of the modulating binary waveform. In the case of digital modulation, the carrier f_c is never transmitted. Rather, one of two frequencies is transmitted, either $f_c + \Delta f$ or $f_c - \Delta f$ depending on the binary state 0 or 1 of the modulation signal.

If $h \ll 1$, the modulation is called *wideband FM* and its bandwidth is approximately $2f_\Delta$. While wideband FM uses more bandwidth, it can improve the signal-to-noise ratio significantly; for example, doubling the value of , while keeping f_m constant, results in an eight-fold improvement in the signal-to-noise ratio. (Compare this with chirp spread spectrum, which uses extremely wide frequency deviations to achieve processing gains comparable to traditional, better-known spread-spectrum modes).

With a tone-modulated FM wave, if the modulation frequency is held constant and the modulation index is increased, the (non-negligible) bandwidth of the FM signal increases but the spacing between spectra remains the same; some spectral components decrease in strength as others increase. If the frequency deviation is held constant and the modulation frequency increased, the spacing between spectra increases.

Frequency modulation can be classified as narrowband if the change in the carrier frequency is about the same as the signal frequency, or as wideband if the change in the carrier frequency is much higher (modulation index > 1) than the signal frequency. For example, narrowband FM (NFM) is used for two-way radio systems such as Family Radio Service, in which the carrier is allowed to deviate only 2.5 kHz above and below the center frequency with speech signals of no more than 3.5 kHz bandwidth. Wideband FM is used for FM broadcasting, in which music and speech are transmitted with up to 75 kHz deviation from the center frequency and carry audio with up to a 20 kHz bandwidth and subcarriers up to 92 kHz.

Bessel Functions

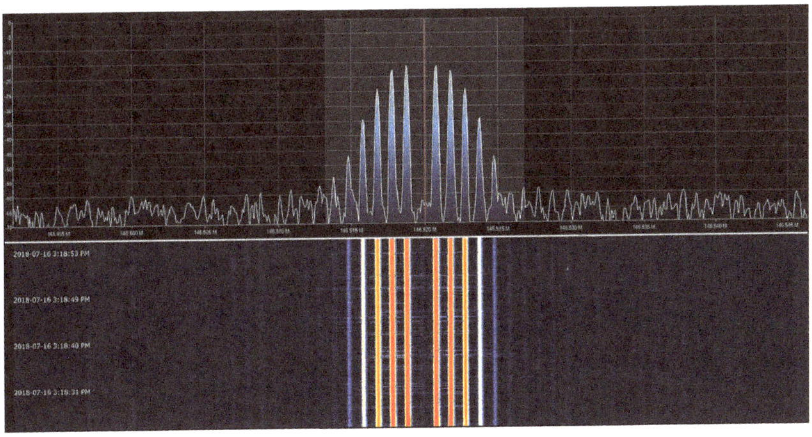

Frequency spectrum and waterfall plot of a 146.52 MHz carrier, frequency modulated by a 1,000 Hz sinusoid. The modulation index has been adjusted to around 2.4, so the carrier frequency has small amplitude. Several strong sidebands are apparent; in principle an infinite number are produced in FM but the higher-order sidebands are of negligible magnitude.

For the case of a carrier modulated by a single sine wave, the resulting frequency spectrum can be calculated using Bessel functions of the first kind, as a function of the sideband number and the modulation index. The carrier and sideband amplitudes are illustrated for different modulation indices of FM signals. For particular values of the modulation index, the carrier amplitude becomes zero and all the signal power is in the sidebands.

Since the sidebands are on both sides of the carrier, their count is doubled, and then multiplied by the modulating frequency to find the bandwidth. For example, 3 kHz deviation modulated by a 2.2 kHz audio tone produces a modulation index of 1.36. Suppose that we limit ourselves to only those sidebands that have relative amplitude of at least 0.01. Then, examining the chart shows this modulation index will produce three sidebands. These three sidebands, when doubled, gives us (6 × 2.2 kHz) or a 13.2 kHz required bandwidth.

Carson's Rule

A rule of thumb, Carson's rule states that nearly all (~98 percent) of the power of a frequency-modulated signal lies within a bandwidth B_T of:

$$B_T = 2(\Delta f + f_m) = 2f_m(\beta + 1)$$

where Δf, as defined above, is the peak deviation of the instantaneous frequency $f(t)$ from the center carrier frequency f_c, β is the Modulation index which is the ratio of frequency deviation to highest frequency in the modulating signal and f_m is the highest frequency in the modulating signal. Condition for application of Carson's rule is only sinusoidal signals.

$$B_T = 2(\Delta f + W) = 2W(D + 1)$$

where W is the highest frequency in the modulating signal but non-sinusoidal in nature and D is the Deviation ratio which the ratio of frequency deviation to highest frequency of modulating non-sinusoidal signal.

Noise Reduction

FM provides improved signal-to-noise ratio (SNR), as compared for example with AM. Compared with an optimum AM scheme, FM typically has poorer SNR below a certain signal level called the noise threshold, but above a higher level – the full improvement or full quieting threshold – the SNR is much improved over AM. The improvement depends on modulation level and deviation. For typical voice communications channels, improvements are typically 5–15 dB. FM broadcasting using wider deviation can achieve even greater improvements. Additional techniques, such as pre-emphasis of higher audio frequencies with corresponding de-emphasis in the receiver, are generally used to improve overall SNR in FM circuits. Since FM signals have constant amplitude, FM receivers normally have limiters that remove AM noise, further improving SNR.

Implementation

Modulation

FM signals can be generated using either direct or indirect frequency modulation:

- Direct FM modulation can be achieved by directly feeding the message into the input of a voltage-controlled oscillator.
- For indirect FM modulation, the message signal is integrated to generate a phase-modulated signal. This is used to modulate a crystal-controlled oscillator, and the result is passed through a frequency multiplier to produce an FM signal. In this modulation, narrowband FM is generated leading to wideband FM later and hence the modulation is known as indirect FM modulation.

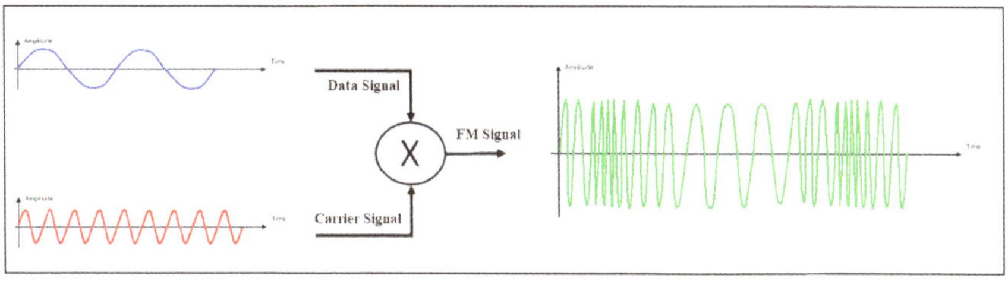

FM Modulation.

Demodulation

Many FM detector circuits exist. A common method for recovering the information signal is through a Foster-Seeley discriminator or ratio detector. A phase-locked loop can be used as an FM demodulator. Slope detection demodulates an FM signal by using a tuned circuit which has its resonant frequency slightly offset from the carrier. As the frequency rises and falls the tuned circuit provides a changing amplitude of response, converting FM to AM. AM receivers may detect some FM transmissions by this means, although it does not provide an efficient means of detection for FM broadcasts.

Applications

Magnetic Tape Storage

FM is also used at intermediate frequencies by analog VCR systems (including VHS) to record the luminance (black and white) portions of the video signal. Commonly, the chrominance component is recorded as a conventional AM signal, using the higher-frequency FM signal as bias. FM is the only feasible method of recording the luminance ("black and white") component of video to (and retrieving video from) magnetic tape without distortion; video signals have a large range of frequency components – from a few hertz to several megahertz, too wide for equalizers to work with due to electronic noise below −60 dB. FM also keeps the tape at saturation level, acting as a form of noise reduction; a limiter can mask variations in playback output, and the FM capture effect removes print-through and pre-echo. A continuous pilot-tone, if added to the signal – as was done on V2000 and many Hi-band formats – can keep mechanical jitter under control and assist timebase correction.

These FM systems are unusual, in that they have a ratio of carrier to maximum modulation frequency of less than two; contrast this with FM audio broadcasting, where the ratio is around 10,000. Consider, for example, a 6-MHz carrier modulated at a 3.5-MHz rate; by Bessel analysis, the first sidebands are on 9.5 and 2.5 MHz and the second sidebands are on 13 MHz and −1 MHz. The result is a reversed-phase sideband on +1 MHz; on demodulation, this results in unwanted output at 6 – 1 = 5 MHz. The system must be designed so that this unwanted output is reduced to an acceptable level.

Sound

FM is also used at audio frequencies to synthesize sound. This technique, known as FM synthesis, was popularized by early digital synthesizers and became a standard feature in several generations of personal computer sound cards.

Radio

An American FM radio transmitter.

Edwin Howard Armstrong was an American electrical engineer who invented wideband frequency modulation (FM) radio. He patented the regenerative circuit in 1914, the superheterodyne receiver in 1918 and the super-regenerative circuit in 1922. Armstrong presented his paper, "A Method of Reducing Disturbances in Radio Signaling by a System of Frequency Modulation", (which first described FM radio) before the New York section of the Institute of Radio Engineers on November 6, 1935. The paper was published in 1936.

As the name implies, wideband FM (WFM) requires a wider signal bandwidth than amplitude modulation by an equivalent modulating signal; this also makes the signal more robust against noise and interference. Frequency modulation is also more robust against signal-amplitude-fading phenomena. As a result, FM was chosen as the modulation standard for high frequency, high fidelity radio transmission, hence the term "FM radio" (although for many years the BBC called it "VHF radio" because commercial FM broadcasting uses part of the VHF band—the FM broadcast band). FM receivers employ a special detector for FM signals and exhibit a phenomenon known as the capture effect, in which the tuner "captures" the stronger of two stations on the same frequency while rejecting the other (compare this with a similar situation on an AM receiver, where both stations can be heard simultaneously). However, frequency drift or a lack of selectivity may cause one station to be overtaken by another on an adjacent channel. Frequency drift was a problem in early (or inexpensive) receivers; inadequate selectivity may affect any tuner.

An FM signal can also be used to carry a stereo signal; this is done with multiplexing and demultiplexing before and after the FM process. The FM modulation and demodulation process is identical in stereo and monaural processes. A high-efficiency radio-frequency switching amplifier can be used to transmit FM signals (and other constant-amplitude signals). For a given signal strength (measured at the receiver antenna), switching amplifiers use less battery power and typically cost less than a linear amplifier. This gives FM another advantage over other modulation methods requiring linear amplifiers, such as AM and QAM.

FM is commonly used at VHF radio frequencies for high-fidelity broadcasts of music and speech. Analog TV sound is also broadcast using FM. Narrowband FM is used for voice communications in commercial and amateur radio settings. In broadcast services, where audio fidelity is important, wideband FM is generally used. In two-way radio, narrowband FM (NBFM) is used to conserve bandwidth for land mobile, marine mobile and other radio services.

There are reports that on October 5, 1924, Professor Mikhail A. Bonch-Bruevich, during a scientific and technical conversation in the Nizhny Novgorod Radio Laboratory, reported about his new method of telephony, based on a change in the period of oscillations. Demonstration of frequency modulation was carried out on the laboratory model.

Phase Modulation (PM)

Phase modulation is defined as the process in which the instantaneous phase of the carrier signal is varied in accordance with the instantaneous amplitude of the modulating signal. In this type of modulation, the amplitude and frequency of the carrier signal remains unaltered after PM. The modulating, signal is mapped to the carrier signal in the form of variations in the instantaneous phase of the carrier signal.

The physical appearance of a PM signal for a sinusoidal modulating signal cannot be illustrated because it is impossible to show the phase changes in the modulated earlier at each instant of time. Therefore, a square wave is used as the modulating signal to show the phase change of the modulated carrier with the change in the modulating amplitude. The modulated carrier signal for a non-sinusoidal, square wave-modulating signal, in this modulating; signal changes its amplitudes at time instances, t1, t2, t3, t4 and t5. According to the definition of PM, the phase of the carrier changes in proportion to the variation in the modulating amplitude at these time instants. Before t1, between time zero and t1, the modulating amplitude is zero. Therefore, during this period, there is no modulation in the Carrier signal and the phase remains unchanged. The same thing happens alter time t5, as the modulating amplitude becomes zero.

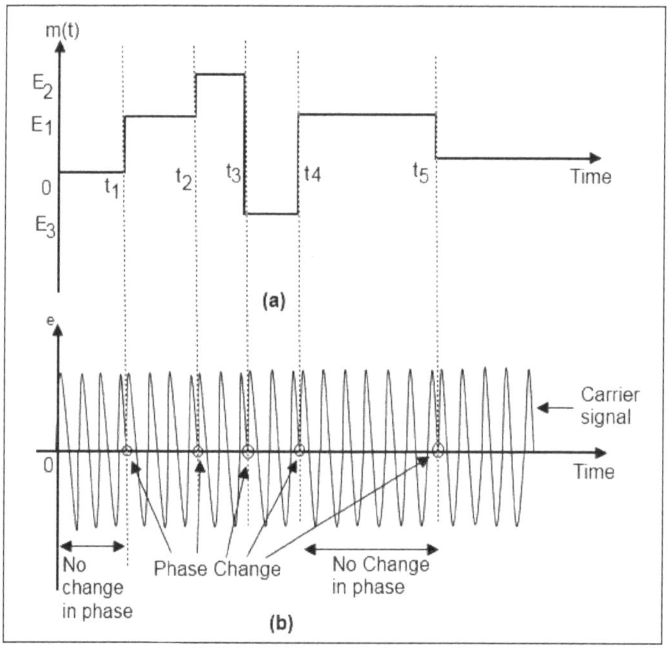

Phase modulation waveforms.

At time t1, the amplitude of m(t) increases from zero to E1. Therefore, at t1, the phase modulated carrier also changes corresponding to E1. This phase remains to this attained value until time t2, as between t1 and t2, the amplitude of m(t) remains constant at El. At t2, the amplitude of m(t) shoots up to E2, and therefore the phase of the carrier again increases corresponding to the increase in m(t). This new value of the phase attained at time t2 remains constant up to time t3. At time t3, m(t) goes negative and its amplitude becomes E3. Consequently, the phase of the carrier also changes and it decreases from the previous value attained at t2. The decrease in phase corresponds to the decrease in amplitude of m(t). The phase of the carrier remains constant during the time interval between t3 and t4. At t4, m(t) goes positive to reach the amplitude El resulting in a corresponding increase in the phase of modulated carrier at time t4. Between t4 and t5, the phase remains constant. At t5 it decreases to the phase of the unmodulated carrier, as the amplitude of m(t) is zero beyond t5.

Equation of a PM Wave

To derive the equation of a PM wave, it is convenient to consider the modulating signal as a pure sinusoidal wave. The carrier signal is always a high frequency sinusoidal wave. Consider the modulating signal, em and the carrier signal:

$$e_m = E_m \cos \omega_m t$$

$$e_c = E_c \sin \omega_c t$$

The initial phases of the modulating signal and the carrier signal are ignored in above equations because they do not contribute to the modulation process due to their constant values. After PM, the phase of the carrier will not remain constant. It will vary according to the modulating signal em maintaining the amplitude and frequency as constants. Suppose, after PM, the equation of the carrier is represented as:

$$e = E_c \sin \theta$$

Where θ, is the instantaneous phase of the modulated carrier, and sinusoid ally varies in proportion to the modulating signal. Therefore, after PM, the instantaneous phase of the modulated carrier can be written as:

$$\theta = w_c t + K_p e_m$$

Where, kp is the constant of proportionality for phase modulation.

Substituting $e_m = E_m \cos \omega t$ in $\theta = w_c t + K_p e_m$, you get:

$$\theta = w_c t + K_p E_m \cos \omega_m t$$

Above, the factor, $K_p E_m$ is defined as the modulation index, and is given as:

$$m_p = K_p E_m$$

where, the subscript p signifies; that mp is the modulation index of the PM wave. Therefore, $\theta = w_c t + K_p E_m \cos \omega_m t$ becomes:

$$\theta = w_c t + m_p \cos \omega_m t$$

Substituting $\theta = w_c t + m_p \cos \omega_m t$ and $e = E_c \sin \theta$, you get:

$$e = E_c \sin(\omega_c t + m_p \cos \omega_m t)$$

This is the final expression of the PM wave.

References

- Analog-communication-introduction, analog-communication: tutorialspoint.com, Retrieved 15 April, 2020
- Telecommunication System Engineering By Roger L. Freeman.2004 John Wiley and Sons. ISBN 0-471-45133-9
- RF-wireless-communication-Analog-VS-Digital-5787: dealna.com, Retrieved 19 January, 2020
- Phase-modulation, electronics: daenotes.com, Retrieved 28 May, 2020
- Principles-of-communication-angle-modulation: tutorialspoint.com, Retrieved 26 February, 2020
- A.P.Godse and U.A.Bakshi (2009). Communication Engineering. Technical Publications. p. 36. ISBN 978-81-8431-089-4

5

Digital Communication and Modulation

The mode of communication in which the information is encoded in a digital format and transferred electronically is known as digital communication. Amplitude shift keying, frequency shift keying, minimum shift keying, phase shift keying, etc. are some of the techniques used in digital transmission and modulation. This chapter discusses the related aspects of digital communication in detail.

Digital communication is the process of devices communicating information digitally. Basic Components of a Digital Communication System are, broadly, every digital Communication system consists of these basic components.

- Source.
- Input transducer.
- Analog to digital converter.
- Source encoder.
- Channel encoder.
- Digital modulator.
- Communication channel.
- Digital demodulator.
- Digital to analog converter.
- Channel decoder.
- Source decoder.
- Output transducer.
- Output signal.

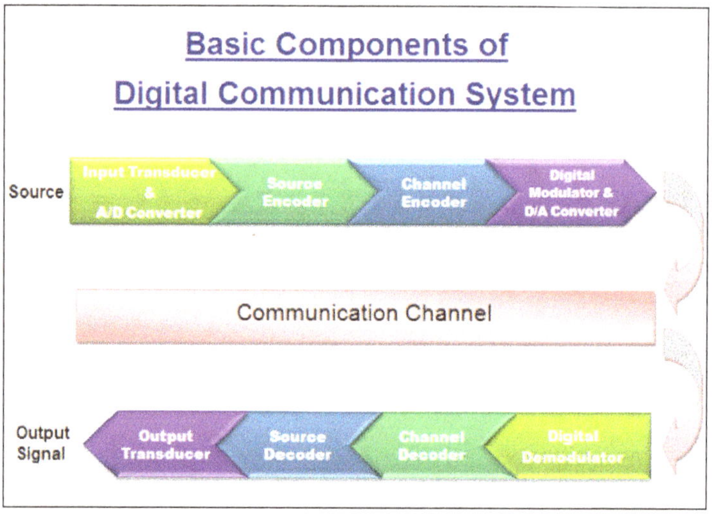

Basic components of digital communication system.

Working Process of Digital Communication

Source

The source consists of an analog signal. For example: A Sound signal.

Input Transducer

This block consists of input transducer which takes a physical input and converts it to an electrical signal For example: Microphone.

Analog to Digital Converter

This electrical signal from Input Transducer is further processed and converted into Digital Signal by Analog to Digital Converter.

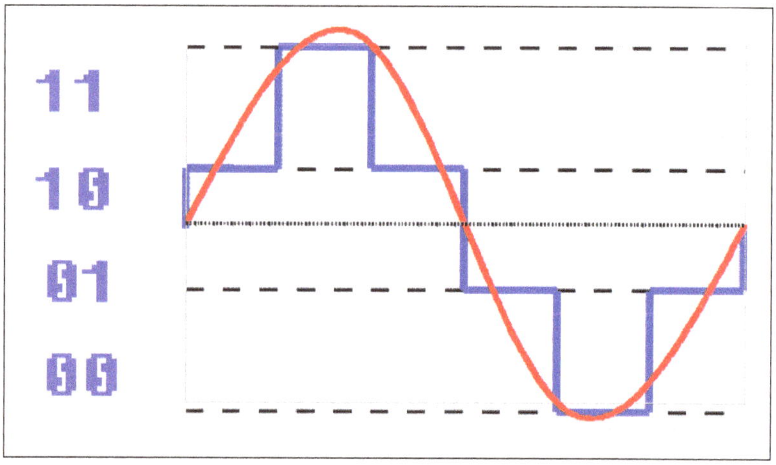

Analog to digital conversion.

Source Encoder

The source encoder compresses the data into lowest number of bits. This procedure helps in efficient operation of the bandwidth. It removes the unnecessary bits.

Channel Encoder

The channel encoder, here the coding is done for error correction. During the transmission of the signal, due to the sound in the channel, the signal may get distorted. To avoid this, the channel encoder adds some unnecessary bits to the transmitted data. These bits are the error correcting bits.

Digital Modulator

Here the signal which is to be transmitted is modulated by a carrier. The carrier is used for for effective long distance transmission of data.

Digital to Analog Converter

The digital signal extracted from the carrier is then converted again into analog so that the signal can be passed effectively through the channel or medium.

Channel

The channel provides a path for the signal and permits the analog signal to transmit from the transmitter end to the receiver end.

Digital Demodulator

This is the place from where the data retrieving process is started at the receiver end. The received signal is demodulated and again converted from analog to digital. The signal gets rebuild here.

Channel Decoder

The channel decoder does the error corrections post sequence detection. The distortions which might take place during the transmission are corrected by adding some additional bits. Addition of these bits help in the complete recovery of the original signal.

Source Decoder

The resulting signal is again digitized by sampling and quantizing. This is done to obtain the unadulterated digital output without any loss of information. The source decoder creates again the source output.

Output Transducer

This is the final block which converts the signal into its original form (which was at the input of the transmitter). It converts the electrical signal into physical output.

For example: Speaker.

Output Signal

This is the output for which the whole process is done. For example: The sound signal received.

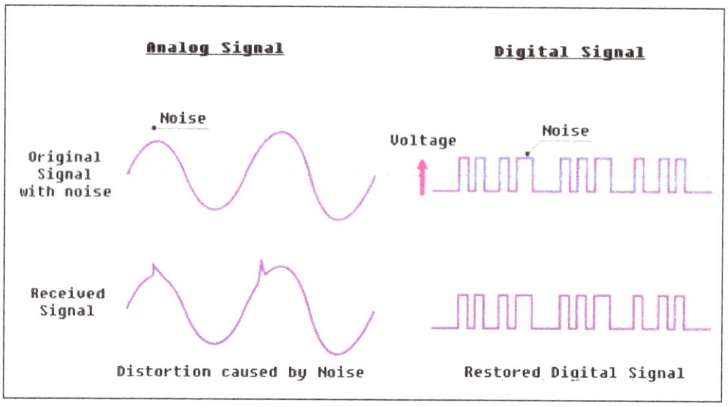

Advantage of digital signal.

Advantages of Digital Communication over Analog Communication

Digital Communication has many advantages over Analog Communication. Some of them are listed below:

- The specific signal level of the digital signal is not very important. Due to this, digital signals are quite unaffected by the flaws of electronic systems that may spoil analog signals.
- The configuration process of digital signals is easier than analog signals.
- Encryption works better in Digital Signals (using codes).
- Digital circuits are more consistent and reliable.
- Digital circuits are easy to design (normally).
- The cost of manufacturing Digital Circuits is lesser than Analog Circuits.
- Digitals Signals do not get corrupted by noise, interference, and distortions.
- Cross-talking is very rare in Digital Communication.

- Long distance data transmission is more easy and cheap with Digital Signals.
- The hardware implementation in digital circuits is much more flexible if compared to analog circuits.
- The method of combining digital signals using Time Division Multiplexing (TDM) is easier than the method of combining analog signals using Frequency Division Multiplexing (FDM).
- Digital signals can be saved and extracted more easily than analog signals.
- Most of the digital circuits have almost common encoding techniques and therefore similar devices can be used for a number of purposes.
- Digital Communication supports multi-dimensional transmissions simultaneously.
- The capability of the channel is efficiently utilized by digital signals.
- The signal is unchanged as the pulse needs a high interruption to change its properties, which is very complex.

Advantages of Digital Communication

As the signals are digitized, there are many advantages of digital communication over analog communication, such as:

- The effect of distortion, noise, and interference is much less in digital signals as they are less affected.
- Digital circuits are more reliable.
- Digital circuits are easy to design and cheaper than analog circuits.
- The hardware implementation in digital circuits, is more flexible than analog.
- The occurrence of cross-talk is very rare in digital communication.
- The signal is un-altered as the pulse needs a high disturbance to alter its properties, which is very difficult.
- Signal processing functions such as encryption and compression are employed in digital circuits to maintain the secrecy of the information.
- The probability of error occurrence is reduced by employing error detecting and error correcting codes.
- Spread spectrum technique is used to avoid signal jamming.
- Combining digital signals using Time Division Multiplexing (TDM) is easier than combining analog signals using Frequency Division Multiplexing (FDM).

- The configuring process of digital signals is easier than analog signals.
- Digital signals can be saved and retrieved more conveniently than analog signals.
- Many of the digital circuits have almost common encoding techniques and hence similar devices can be used for a number of purposes.
- The capacity of the channel is effectively utilized by digital signals.

Digital Transmission

Data transmission (also data communication or digital communications) is the transfer of data (a digital bitstream or a digitized analog signal) over a point-to-point or point-to-multipoint communication channel. Examples of such channels are copper wires, optical fibers, wireless communication channels, storage media and computer buses. The data are represented as an electromagnetic signal, such as an electrical voltage, radiowave, microwave, or infrared signal.

Analog or analogue transmission is a transmission method of conveying voice, data, image, signal or video information using a continuous signal which varies in amplitude, phase, or some other property in proportion to that of a variable. The messages are either represented by a sequence of pulses by means of a line code (baseband transmission), or by a limited set of continuously varying wave forms (passband transmission), using a digital modulation method. The passband modulation and corresponding demodulation (also known as detection) is carried out by modem equipment. According to the most common definition of digital signal, both baseband and passband signals representing bit-streams are considered as digital transmission, while an alternative definition only considers the baseband signal as digital, and passband transmission of digital data as a form of digital-to-analog conversion.

Data transmitted may be digital messages originating from a data source, for example a computer or a keyboard. It may also be an analog signal such as a phone call or a video signal, digitized into a bit-stream, for example, using pulse-code modulation (PCM) or more advanced source coding (analog-to-digital conversion and data compression) schemes. This source coding and decoding is carried out by codec equipment.

Protocol Layers and Sub-topics

Courses and textbooks in the field of data transmission typically deal with the following OSI model protocol layers and topics:

- Layer 1, the physical layer:
  - Channel coding including:
    - Digital modulation schemes.

        - Line coding schemes.
        - Forward error correction (FEC) codes.
    - Bit synchronization.
    - Multiplexing.
    - Equalization.
    - Channel models.
- Layer 2, the data link layer:
    - Channel access schemes, media access control (MAC).
    - Packet mode communication and Frame synchronization.
    - Error detection and automatic repeat request (ARQ).
    - Flow control.
- Layer 6, the presentation layer:
    - Source coding (digitization and data compression), and information theory.
    - Cryptography (may occur at any layer).

Applications

Data (mainly but not exclusively informational) has been sent via non-electronic (e.g. optical, acoustic, mechanical) means since the advent of communication. Analog signal data has been sent electronically since the advent of the telephone. However, the first data electromagnetic transmission applications in modern time were telegraphy and teletypewriters which are both digital signals. The fundamental theoretical work in data transmission and information theory by Harry Nyquist, Ralph Hartley, Claude Shannon and others during the early 20th century, was done with these applications in mind.

Data transmission is utilized in computers in computer buses and for communication with peripheral equipment via parallel ports and serial ports such as RS-232, Firewire and USB. The principles of data transmission are also utilized in storage media for Error detection and correction since 1951.

Data transmission is utilized in computer networking equipment such as modems, local area networks (LAN) adapters, repeaters, repeater hubs, microwave links, wireless network access points, etc.

In telephone networks, digital communication is utilized for transferring many phone calls over the same copper cable or fiber cable by means of Pulse code modulation (PCM), i.e. sampling and digitization, in combination with Time division multiplexing (TDM. Telephone exchanges have become digital and software controlled, facilitating many value added services. For example, the first AXE telephone exchange was presented in 1976. Since the late 1980s, digital communication to the end user has been possible using Integrated Services Digital Network (ISDN) services. Since the end of the 1990s, broadband access techniques such as ADSL, Cable modems, fiber-to-the-building (FTTB) and fiber-to-the-home (FTTH) have become widespread to small offices and homes. The current tendency is to replace traditional telecommunication services by packet mode communication such as IP telephony and IPTV.

Transmitting analog signals digitally allows for greater signal processing capability. The ability to process a communications signal means that errors caused by random processes can be detected and corrected. Digital signals can also be sampled instead of continuously monitored. The multiplexing of multiple digital signals is much simpler to the multiplexing of analog signals.

Because of all these advantages, and because recent advances in wideband communication channels and solid-state electronics have allowed scientists to fully realize these advantages, digital communications has grown quickly. Digital communications is quickly edging out analog communication because of the vast demand to transmit computer data and the ability of digital communications to do so.

The digital revolution has also resulted in many digital telecommunication applications where the principles of data transmission are applied. Examples are second-generation and later cellular telephony, video conferencing, digital TV, digital radio, telemetry, etc.

Data transmission, digital transmission or digital communications is the physical transfer of data (a digital bit stream or a digitized analog signal) over a point-to-point or point-to-multipoint communication channel. Examples of such channels are copper wires, optical fibers, wireless communication channels, storage media and computer buses. The data are represented as an eectromagnetic signal, such as an electrical voltage, radiowave, microwave, or infrared signal.

While analog transmission is the transfer of a continuously varying analog signal over an analog channel, digital communications is the transfer of discrete messages over a digital or an analog channel. The messages are either represented by a sequence of pulses by means of a line code (baseband transmission), or by a limited set of continuously varying wave forms (passband transmission), using a digital modulation method. The passband modulation and corresponding demodulation (also known as detection) is carried out by modem equipment. According to the most common definition of digital signal, both baseband and passband signals representing bit-streams are considered as digital transmission, while an alternative definition only considers the baseband signal as digital, and passband transmission of digital data as a form of digital-to-analog conversion.

Data transmitted may be digital messages originating from a data source, for example a computer or a keyboard. It may also be an analog signal such as a phone call or a video signal, digitized into a bit-stream for example using pulse-code modulation (PCM) or more advanced source coding (analog-to-digital conversion and data compression) schemes. This source coding and decoding is carried out by codec equipment.

Serial and Parallel Transmission

In telecommunications, serial transmission is the sequential transmission of signal elements of a group representing a character or other entity of data. Digital serial transmissions are bits sent over a single wire, frequency or optical path sequentially. Because it requires less signal processing and fewer chances for error than parallel transmission, the transfer rate of each individual path may be faster. This can be used over longer distances as a check digit or parity bit can be sent along it easily.

In telecommunications, parallel transmission is the simultaneous transmission of the signal elements of a character or other entity of data. In digital communications, parallel transmission is the simultaneous transmission of related signal elements over two or more separate paths. Multiple electrical wires are used which can transmit multiple bits simultaneously, which allows for higher data transfer rates than can be achieved with serial transmission. This method is used internally within the computer, for example the internal buses, and sometimes externally for such things as printers, The major issue with this is "skewing" because the wires in parallel data transmission have slightly different properties (not intentionally) so some bits may arrive before others, which may corrupt the message. A parity bit can help to reduce this. However, electrical wire parallel data transmission is therefore less reliable for long distances because corrupt transmissions are far more likely.

Communication Channels

Some communications channel types include:

- Data transmission circuit.
- Full-duplex.
- Half-duplex.
- Multi-drop:
    - Bus network.
    - Mesh network.
    - Ring network.
    - Star network.

 - Wireless network.
- Point-to-point.
- Simplex.

Asynchronous and Synchronous Data Transmission

Asynchronous serial communication uses start and stop bits to signify the beginning and end of transmission. This method of transmission is used when data are sent intermittently as opposed to in a solid stream.

Synchronous transmission synchronizes transmission speeds at both the receiving and sending end of the transmission using clock signals. The clock may be a separate signal or embedded in the data. A continual stream of data is then sent between the two nodes. Due to there being no start and stop bits the data transfer rate is more efficient.

Digital Modulation

DM stands for Digital Modulation and is a generic name for modulation techniques that uses discrete signals to modulate a carrier wave. In comparison, FM and AM are analog techniques. The three main types of digital modulation are Frequency Shift Keying (FSK), Phase Shift Keying (PSK) and Amplitude Shift Keying (ASK). DM eliminates transmission noise and offers improved robustness to signal interference. However, it is not uncommon for DM to introduce time delay due to the processing required. Digital Modulation provides more information capacity, high data security, quicker system availability with great quality communication. Hence, digital modulation techniques have a greater demand, for their capacity to convey larger amounts of data than analog modulation techniques. There are many types of digital modulation techniques and also their combinations, depending upon the need.

Constellation Diagram

A constellation diagram is a representation of a signal modulated by a digital modulation scheme such as quadrature amplitude modulation or phase-shift keying. It displays the signal as a two-dimensional xy-plane scatter diagram in the complex plane at symbol sampling instants. The angle of a point, measured counterclockwise from the horizontal axis, represents the phase shift of the carrier wave from a reference phase. The distance of a point from the origin represents a measure of the amplitude or power of the signal.

In a digital modulation system, information is transmitted as a series of samples, each occupying a uniform time slot. During each sample, the carrier wave has a constant amplitude and phase, which is restricted to one of a finite number of values. So each sample encodes one of a finite number of “symbols”, which in turn represent one or more binary digits (bits) of information. Each symbol is encoded as a different combination of amplitude and phase of the carrier, so each symbol is represented by a point on the constellation diagram, called a constellation point. The constellation diagram shows all the possible symbols that can be transmitted by the system as a collection of points. In a frequency or phase modulated signal, the signal amplitude is constant, so the points lie on a circle around the origin.

The carrier representing each symbol can be created by adding together different amounts of a cosine wave representing the "I" or in-phase carrier, and a sine wave, shifted by 90° from the I carrier called the "Q" or quadrature carrier. Thus each symbol can be represented by a complex number, and the constellation diagram can be regarded as a complex plane, with the horizontal real axis representing the I component and the vertical imaginary axis representing the Q component. A coherent detector is able to independently demodulate these carriers. This principle of using two independently modulated carriers is the foundation of quadrature modulation. In pure phase modulation, the phase of the modulating symbol is the phase of the carrier itself and this is the best representation of the modulated signal.

A 'signal space diagram' is an ideal constellation diagram showing the correct position of the point representing each symbol. After passing through a communication channel, due to electronic noise or distortion added to the signal, the amplitude and phase received by the demodulator may differ from the correct value for the symbol. When plotted on a constellation diagram the point representing that received sample will be offset from the correct position for that symbol. An electronic test instrument called a vector signal analyzer can display the constellation diagram of a digital signal by sampling the signal and plotting each received symbol as a point. The result is a 'ball' or 'cloud' of points surrounding each symbol position. Measured constellation diagrams can be used to recognize the type of interference and distortion in a signal.

Interpretation

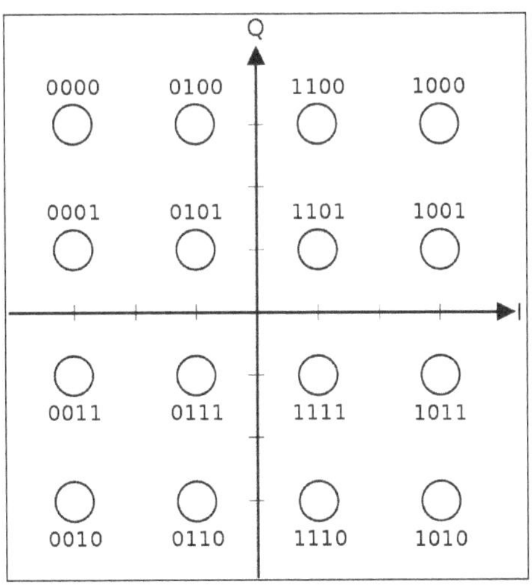

A constellation diagram for rectangular 16-QAM.

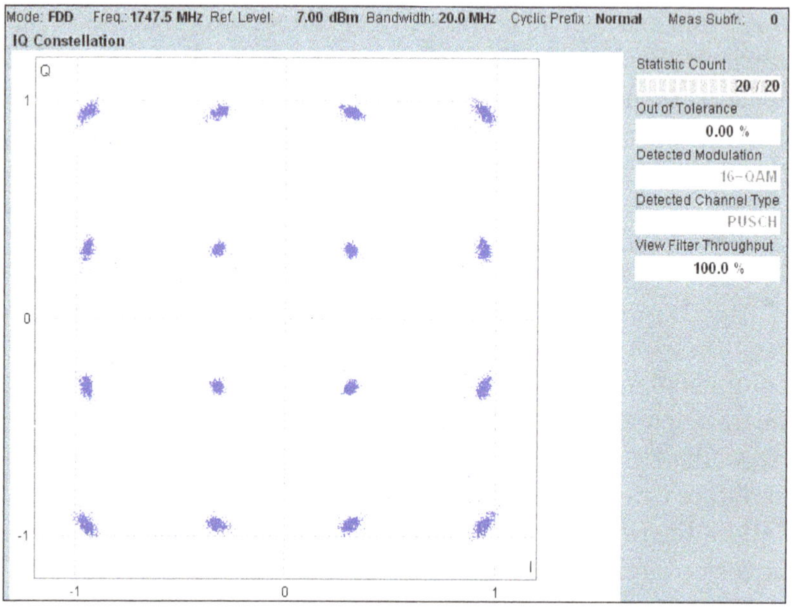

The constellation as received, with noise added.

The number of constellation points in a diagram gives the size of the "alphabet" of symbols that can be transmitted by each sample, and so determines the number of bits transmitted per sample. It is usually a power of 2. A diagram with four points, for example, represents a modulation scheme that can separately encode all 4 combinations of two bits: 00, 01, 10, and 11 and so can transmit two bits per sample. Thus in general a modulation with N constellation points transmits $\log_2 N$ bits per sample.

After passing through the communication channel the signal is decoded by a demodulator. The function of the demodulator is to classify each sample as a symbol. The set of sample values which the demodulator classifies as a given symbol can be represented by a region in the plane drawn around each constellation point. If noise causes the point representing a sample to stray into the region representing another symbol, the demodulator will misidentify that sample as the other symbol, resulting in a symbol error. Most demodulators choose, as its estimate of what was actually transmitted, the constellation point which is closest (in a Euclidean distance sense) to that of the received sample; this is called maximum likelihood detection. On the constellation diagram these detection regions can be easily represented by dividing the plane by lines equidistant from each adjacent pair of points.

One half the distances between each pair of neighboring points is the amplitude of additive noise or distortion required to cause one of the points to be misidentified as the other, and thus cause a symbol error. Therefore, the further the points are separated from one another, the greater the noise immunity of the modulation. Practical modulation systems are designed to maximize the minimum noise needed to cause a symbol error; on the constellation diagram this means that the distance between each pair of adjacent points is equal.

The received signal quality can be analyzed by displaying the constellation diagram of the signal at the receiver on a vector signal analyzer. Some types of distortion show up as characteristic patterns on the diagram:

- Gaussian noise causes the samples to land in a random ball about each constellation point.
- Non-coherent single frequency interference shows as samples making circles about each constellation point.
- Phase noise shows as constellation points spreading into arcs centered on the origin.
- Amplifier compression causes the corner points to move towards the center.

A constellation diagram visualizes phenomena similar to those an eye pattern does for one-dimensional signals. The eye pattern can be used to see timing jitter in one dimension of modulation.

Digital Modulation Techniques

Amplitude Shift Keying

Amplitude-shift keying (ASK) is a form of amplitude modulation that represents digital data as variations in the amplitude of a carrier wave. In an ASK system, the binary symbol 1 is represented by transmitting a fixed-amplitude carrier wave and fixed frequency for a bit duration of T seconds. If the signal value is 1 then the carrier signal will be transmitted; otherwise, a signal value of 0 will be transmitted.

Any digital modulation scheme uses a finite number of distinct signals to represent digital data. ASK uses a finite number of amplitudes, each assigned a unique pattern of binary digits. Usually, each amplitude encodes an equal number of bits. Each pattern of bits forms the symbol that is represented by the particular amplitude. The demodulator, which is designed specifically for the symbol-set used by the modulator, determines the amplitude of the received signal and maps it back to the symbol it represents, thus recovering the original data. Frequency and phase of the carrier are kept constant.

Like AM, an ASK is also linear and sensitive to atmospheric noise, distortions, propagation conditions on different routes in PSTN, etc. Both ASK modulation and demodulation processes are relatively inexpensive. The ASK technique is also commonly used to transmit digital data over optical fiber. For LED transmitters, binary 1 is represented by a short pulse of light and binary 0 by the absence of light. Laser transmitters normally have a fixed "bias" current that causes the device to emit a low light level. This low level represents binary 0, while a higher-amplitude lightwave represents binary 1.

The simplest and most common form of ASK operates as a switch, using the presence of a carrier wave to indicate a binary one and its absence to indicate a binary zero. This type of modulation is called on-off keying (OOK), and is used at radio frequencies to transmit Morse code (referred to as continuous wave operation).

More sophisticated encoding schemes have been developed which represent data in groups using additional amplitude levels. For instance, a four-level encoding scheme can represent two bits with each shift in amplitude; an eight-level scheme can represent three bits; and so on. These forms of amplitude-shift keying require a high signal-to-noise ratio for their recovery, as by their nature much of the signal is transmitted at reduced power.

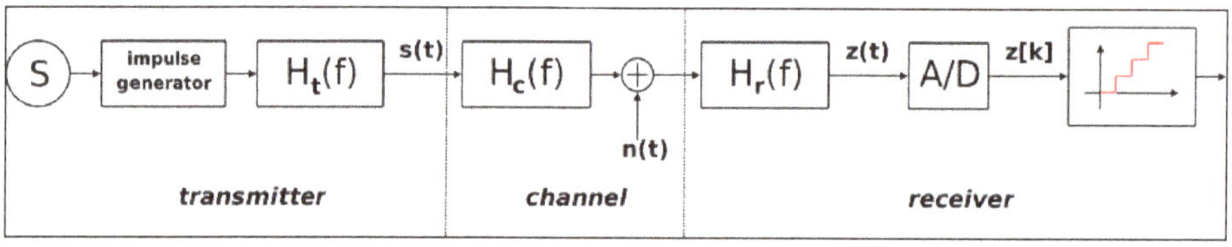

ASK diagram.

ASK system can be divided into three blocks. The first one represents the transmitter, the second one is a linear model of the effects of the channel, and the third one shows the structure of the receiver. The following notation is used:

- $h_t(f)$ is the carrier signal for the transmission.
- $h_c(f)$ is the impulse response of the channel.
- n(t) is the noise introduced by the channel.
- $h_r(f)$ is the filter at the receiver.
- L is the number of levels that are used for transmission.
- T_s is the time between the generation of two symbols.

Different symbols are represented with different voltages. If the maximum allowed value for the voltage is A, then all the possible values are in the range [–A, A] and they are given by:

$$v_i = \frac{2A}{L-1} i - A; \quad i = 0,1,\ldots,L-1.$$

The difference between one voltage and the other is:

$$\Delta = \frac{2A}{L-1}.$$

Considering the picture, the symbols v[n] are generated randomly by the source S, then the impulse generator creates impulses with an area of v[n]. These impulses are sent to the filter h_t to be sent through the channel. In other words, for each symbol a different carrier wave is sent with the relative amplitude.

Out of the transmitter, the signal s(t) can be expressed in the form:

$$s(t) = \sum_{n=-\infty}^{\infty} v[n] \cdot h_t(t - nT_s).$$

In the receiver, after the filtering through hr (t) the signal is:

$$z(t) = n_r(t) + \sum_{n=-\infty}^{\infty} v[n] \cdot g(t - nT_s),$$

where we use the notation:

$$n_r(t) = n(t) * h_r(t)$$
$$g(t) = h_t(t) * h_c(t) * h_r(t),$$

where * indicates the convolution between two signals. After the A/D conversion the signal z[k] can be expressed in the form:

$$z[k] = n_r[k] + v[k]g[0] + \sum_{n \neq k} v[n]g[k-n].$$

In this relationship, the second term represents the symbol to be extracted. The others are unwanted: the first one is the effect of noise, the third one is due to the intersymbol interference.

If the filters are chosen so that g(t) will satisfy the Nyquist ISI criterion, then there will be no intersymbol interference and the value of the sum will be zero, so:

$$z[k] = n_r[k] + v[k]g[0].$$

The transmission will be affected only by noise.

Probability of Error

The probability density function of having an error of a given size can be modelled by a Gaussian function; the mean value will be the relative sent value, and its variance will be given by:

$$\sigma_N^2 = \int_{-\infty}^{+\infty} \Phi_N(f) \cdot |H_r(f)|^2 \, df,$$

where $\Phi_N(f)$ is the spectral density of the noise within the band and Hr (f) is the continuous Fourier transform of the impulse response of the filter hr (f).

The probability of making an error is given by:

$$P_e = P_{e|H_0} \cdot P_{H_0} + P_{e|H_1} \cdot P_{H_1} + \cdots + P_{e|H_{L-1}} \cdot P_{H_{L-1}} = \sum_{k=0}^{L-1} P_{e|H_k} \cdot P_{H_k},$$

where, for example, $P_{e|H_0}$ is the conditional probability of making an error given that a symbol vo has been sent and P_{H_0} is the probability of sending a symbol vo.

If the probability of sending any symbol is the same, then:

$$P_{H_i} = \frac{1}{L}.$$

If we represent all the probability density functions on the same plot against the possible value of the voltage to be transmitted, we get a picture like this (the particular case of $L = 4$ is shown):

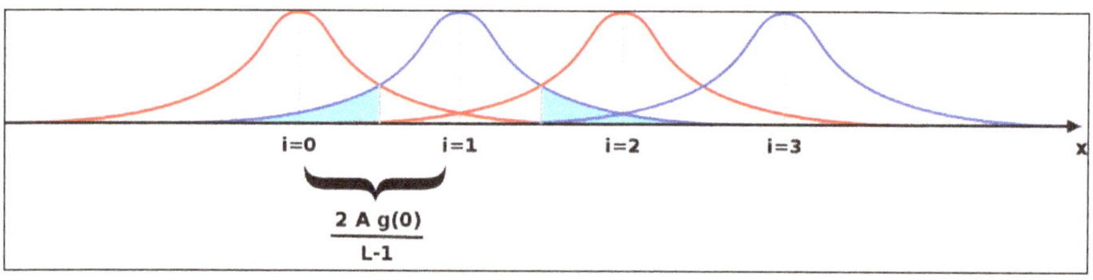

The probability of making an error after a single symbol has been sent is the area of the Gaussian function falling under the functions for the other symbols. It is shown in cyan for just one of them. If we call P^+ the area under one side of the Gaussian, the sum of all the areas will be: $2LP^+ - 2P^+$. The total probability of making an error can be expressed in the form:

$$P_e = 2\left(1 - \frac{1}{L}\right) P^+.$$

We now have to calculate the value of P^+ In order to do that, we can move the origin of the reference wherever we want: The area below the function will not change. We are in a situation like the one shown in the following picture:

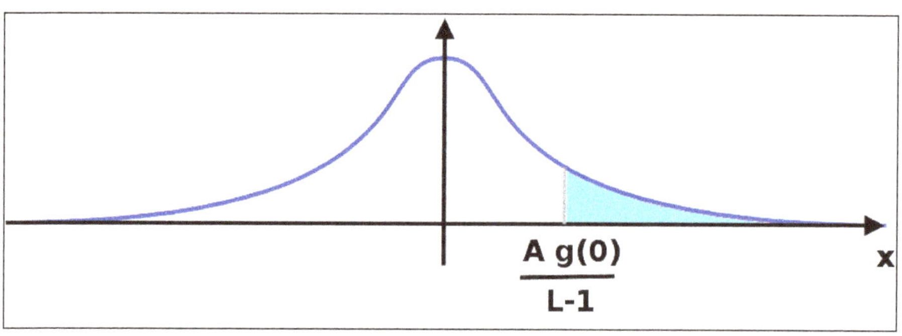

It does not matter which Gaussian function we are considering, the area we want to calculate will be the same. The value we are looking for will be given by the following integral:

$$P^{+} = \int_{\frac{Ag(0)}{L-1}}^{\infty} \frac{1}{\sqrt{2\pi}\sigma_N} e^{-\frac{x^2}{2\sigma_N^2}} dx = \frac{1}{2}\operatorname{erfc}\left(\frac{Ag(0)}{\sqrt{2}(L-1)\sigma_N} \right),$$

where $\operatorname{erfc}(x)$ is the complementary error function. Putting all these results together, the probability to make an error is:

$$P_e = \left(1 - \frac{1}{L}\right)\operatorname{erfc}\left(\frac{Ag(0)}{\sqrt{2}(L-1)\sigma_N} \right).$$

From this formula we can easily understand that the probability to make an error decreases if the maximum amplitude of the transmitted signal or the amplification of the system becomes greater; on the other hand, it increases if the number of levels or the power of noise becomes greater.

This relationship is valid when there is no intersymbol interference, i.e. $g(t)$ is a Nyquist function.

Frequency Shift Keying

Frequency-shift keying (FSK) is a frequency modulation scheme in which digital information is transmitted through discrete frequency changes of a carrier signal. The technology is used for communication systems such as telemetry, weather balloon radiosondes, caller ID, garage door openers, and low frequency radio transmission in the VLF and ELF bands. The simplest FSK is binary FSK (BFSK). BFSK uses a pair of discrete frequencies to transmit binary (0s and 1s) information. With this scheme, the "1" is called the mark frequency and the "0" is called the space frequency.

Modulating and Demodulating

Reference implementations of FSK modems exist and are documented in detail.The demodulation of a binary FSK signal can be done using the Goertzel algorithm very efficiently, even on low-power microcontrollers.

Other Forms of FSK

Continuous-phase Frequency-shift Keying

In principle FSK can be implemented by using completely independent free-running oscillators, and switching between them at the beginning of each symbol period. In general, independent oscillators will not be at the same phase and therefore the same amplitude at the switch-over instant, causing sudden discontinuities in the transmitted signal.

In practice, many FSK transmitters use only a single oscillator, and the process of switching to a different frequency at the beginning of each symbol period preserves the phase. The elimination of discontinuities in the phase (and therefore elimination of sudden changes in amplitude) reduces sideband power, reducing interference with neighboring channels.

Gaussian Frequency-shift Keying

Rather than directly modulating the frequency with the digital data symbols, "instantaneously" changing the frequency at the beginning of each symbol period, Gaussian frequency-shift keying (GFSK) filters the data pulses with a Gaussian filter to make the transitions smoother. This filter has the advantage of reducing sideband power, reducing interference with neighboring channels, at the cost of increasing intersymbol interference. It is used by Improved Layer 2 Protocol, DECT, Bluetooth, Cypress WirelessUSB, Nordic Semiconductor, Texas Instruments LPRF, IEEE 802.15.4, Z-Wave and Wavenis devices. For basic data rate Bluetooth the minimum deviation is 115 kHz.

A GFSK modulator differs from a simple frequency-shift keying modulator in that before the baseband waveform (levels −1 and +1) goes into the FSK modulator, it is passed through a Gaussian filter to make the transitions smoother so to limit its spectral width. Gaussian filtering is a standard way for reducing spectral width; it is called "pulse shaping" in this application.

In ordinary non-filtered FSK, at a jump from −1 to +1 or +1 to −1, the modulated waveform changes rapidly, which introduces large out-of-band spectrum. If the pulse is changed going from −1 to +1 as −1, −0.98, −0.93, ..., +0.93, +0.98, +1, and this smoother pulse is used to determine the carrier frequency, the out-of-band spectrum will be reduced.

Minimum-shift Keying

Minimum frequency-shift keying or minimum-shift keying (MSK) is a particular spectrally efficient form of coherent FSK. In MSK, the difference between the higher and lower frequency is identical to half the bit rate. Consequently, the waveforms that represent a 0 and a 1 bit differ by exactly half a carrier period. The maximum frequency deviation is $\delta = 0.25 f_m$, where f_m is the maximum modulating frequency. As a result, the modulation index m is 0.5. This is the smallest FSK modulation index that can be chosen such that the waveforms for 0 and 1 are orthogonal.

Gaussian Minimum-shift Keying

A variant of MSK called Gaussian minimum-shift keying (GMSK) is used in the GSM mobile phone standard.

Audio FSK

Audio frequency-shift keying (AFSK) is a modulation technique by which digital data is represented by changes in the frequency (pitch) of an audio tone, yielding an encoded signal suitable for transmission via radio or telephone. Normally, the transmitted audio alternates between two tones: one, the "mark", represents a binary one; the other, the "space", represents a binary zero.

AFSK differs from regular frequency-shift keying in performing the modulation at baseband frequencies. In radio applications, the AFSK-modulated signal normally is being used to modulate an RF carrier (using a conventional technique, such as AM or FM) for transmission.

AFSK is not always used for high-speed data communications, since it is far less efficient in both power and bandwidth than most other modulation modes. In addition to its simplicity, however, AFSK has the advantage that encoded signals will pass through AC-coupled links, including most equipment originally designed to carry music or speech.

AFSK is used in the U.S.-based Emergency Alert System to notify stations of the type of emergency, locations affected, and the time of issue without actually hearing the text of the alert.

Continuous 4 Level Continuous

Phase 1 radios in the Project 25 system use continuous 4-level FM (C4FM) modulation.

Minimum-shift Keying

In digital modulation, minimum-shift keying (MSK) is a type of continuous-phase frequency-shift keying that was developed in the late 1950s by Collins Radio employees Melvin L. Doelz and Earl T. Heald. Similar to OQPSK, MSK is encoded with bits alternating between quadrature components, with the Q component delayed by half the symbol period.

However, instead of square pulses as OQPSK uses, MSK encodes each bit as a half sinusoid. This results in a constant-modulus signal (constant envelope signal), which reduces problems caused by non-linear distortion. In addition to being viewed as related to OQPSK, MSK can also be viewed as a continuous phase frequency shift keyed (CPFSK) signal with a frequency separation of one-half the bit rate.

In MSK the difference between the higher and lower frequency is identical to half the bit rate. Consequently, the waveforms used to represent a 0 and a 1 bit differ by exactly half a carrier period. Thus, the maximum frequency deviation is $\delta = 0.25 f_m$ where f_m is the maximum modulating frequency. As a result, the modulation index m is 0.5. This is the smallest FSK modulation index that can be chosen such that the waveforms for 0 and 1 are orthogonal. A variant of MSK called Gaussian minimum-shift keying (GMSK) is used in the GSM mobile phone standard.

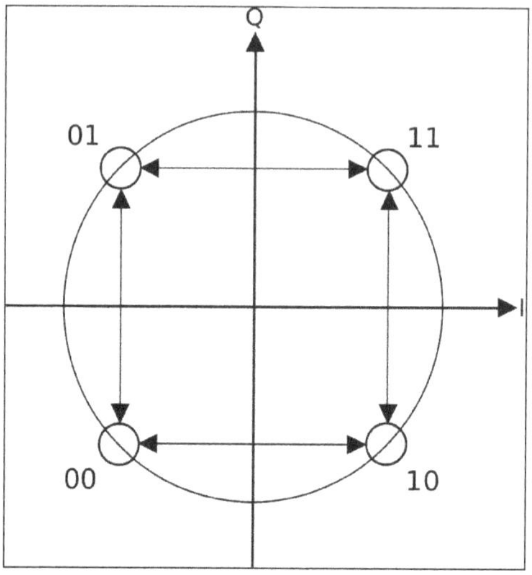

Mapping changes in continuous phase. Each bit time, the carrier phase changes by ±90°

Mathematical Representation

The resulting signal is represented by the formula:

$$s(t) = a_I(t)\cos\left(\frac{\pi t}{2T}\right)\cos(2\pi f_c t) - a_Q(t)\sin\left(\frac{\pi t}{2T}\right)\sin(2\pi f_c t),$$

where $a_I(t)$ and $a_Q(t)$ encode the even and odd information respectively with a sequence of square pulses of duration *2T*. $a_I(t)$ has its pulse edges on $t = [-T, T, 3T, \ldots]$ and $a_Q(t)$ on $t = [0, 2T, 4T, \ldots]$. The carrier frequency is f_c.

Using the trigonometric identity, this can be rewritten in a form where the phase and frequency modulation are more obvious,

$$s(t) = \cos\left[2\pi f_c t + b_k(t)\frac{\pi t}{2T} + \phi_k\right],$$

where $b_k(t)$ is +1 when $a_I(t) = a_Q(t)$ and -1 if they are of opposite signs, and ϕ_k is 0 if $a_I(t)$ is 1, and π otherwise. Therefore, the signal is modulated in frequency and phase, and the phase changes continuously and linearly.

Since the minimum symbol distance is the same as in the QPSK, the following formula can be used for the theoretical bit-error ratio bound:

$$P_b = Q\left(\sqrt{\frac{2E_b}{N_0}}\right) = \frac{1}{2}\operatorname{erfc}\left(\sqrt{\frac{E_b}{N_0}}\right),$$

where E_b is the energy per one bit, N_0 is the noise spectral density, $Q(*)$ denotes the Marcum Q-function and erfc denotes the complementary error function.

Gaussian Minimum-shift Keying

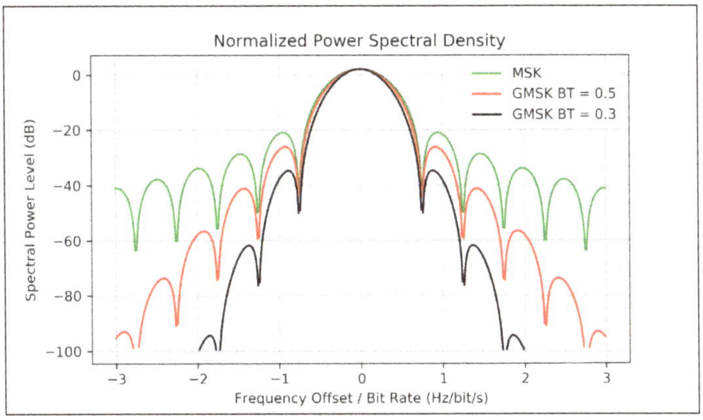

Power spectral densities of the MSK and GMSK. Note that the increasing of time-bandwidth BT negatively influences bit-error-rate performance due to increasing of the ISI.

In digital communication, Gaussian minimum shift keying or GMSK is a continuous-phase frequency-shift keying modulation scheme.

GMSK is similar to standard minimum-shift keying (MSK); however, the digital data stream is first shaped with a Gaussian filter before being applied to a frequency modulator, and typically has much narrower phase shift angles than most MSK modulation systems. This has the advantage of reducing sideband power, which in turn reduces out-of-band interference between signal carriers in adjacent frequency channels.

However, the Gaussian filter increases the modulation memory in the system and causes intersymbol interference, making it more difficult to differentiate between different transmitted data values and requiring more complex channel equalization algorithms such as an adaptive equalizer at the receiver. GMSK has high spectral efficiency, but it needs a higher power level than QPSK, for instance, in order to reliably transmit the same amount of data.

GMSK is most notably used in the Global System for Mobile Communications (GSM) and the satellite communications, e.g., in the Automatic Identification System (AIS) for maritime navigation.

Applications

In 1910, Reginald Fessenden invented a two-tone method of transmitting Morse code. Dots and dashes were replaced with different tones of equal length. The intent was to minimize transmission time.

Some early CW transmitters employed an arc converter that could not be conveniently keyed. Instead of turning the arc on and off, the key slightly changed the transmitter frequency in a technique known as the compensation-wave method. The compensation-wave was not used at the receiver. Spark transmitters used for this method consumed a lot of bandwidth and caused interference, so it was discouraged by 1921.

Most early telephone-line modems used audio frequency-shift keying (AFSK) to send and receive data at rates up to about 1200 bits per second. The Bell 103 and Bell 202 modems used this technique. Even today, North American caller ID uses 1200 baud AFSK in the form of the Bell 202 standard. Some early microcomputers used a specific form of AFSK modulation, the Kansas City standard, to store data on audio cassettes. AFSK is still widely used in amateur radio, as it allows data transmission through unmodified voiceband equipment.

Phase-shift Keying

Phase-shift keying (PSK) is a digital modulation process which conveys data by changing (modulating) the phase of a constant frequency reference signal (the carrier wave). The modulation is accomplished by varying the sine and cosine inputs at a precise time. It is widely used for wireless LANs, RFID and Bluetooth communication.

Any digital modulation scheme uses a finite number of distinct signals to represent digital data. PSK uses a finite number of phases, each assigned a unique pattern of binary digits. Usually, each phase encodes an equal number of bits. Each pattern of bits forms the symbol that is represented by the particular phase. The demodulator, which is designed specifically for the symbol-set used by the modulator, determines the phase of the received signal and maps it back to the symbol it represents, thus recovering the original data. This requires the receiver to be able to compare the phase of the received signal to a reference signal – such a system is termed coherent (and referred to as CPSK).

CPSK requires a complicated demodulator, because it must extract the reference wave from the received signal and keep track of it, to compare each sample to. Alternatively, the phase shift of each symbol sent can be measured with respect to the phase of the previous symbol sent. Because the symbols are encoded in the difference in phase between successive samples, this is called differential phase-shift keying (DPSK). DPSK can be significantly simpler to implement than ordinary PSK, as it is a 'non-coherent' scheme, i.e. there is no need for the demodulator to keep track of a reference wave. A trade-off is that it has more demodulation errors.

Applications

Owing to PSK's simplicity, particularly when compared with its competitor quadrature amplitude modulation, it is widely used in existing technologies.

The wireless LAN standard, IEEE 802.11b-1999, uses a variety of different PSKs depending on the data rate required. At the basic rate of 1 Mbit/s, it uses DBPSK (differential BPSK). To provide the extended rate of 2 Mbit/s, DQPSK is used. In reaching 5.5 Mbit/s and the full rate of 11 Mbit/s, QPSK is employed, but has to be coupled with complementary code keying. The higher-speed wireless LAN standard, IEEE 802.11g-2003, has eight data rates: 6, 9, 12, 18, 24, 36, 48 and 54 Mbit/s. The 6 and 9 Mbit/s modes use OFDM modulation where each sub-carrier is BPSK modulated. The 12 and 18 Mbit/s modes use OFDM with QPSK. The fastest four modes use OFDM with forms of quadrature amplitude modulation.

Because of its simplicity, BPSK is appropriate for low-cost passive transmitters, and is used in RFID standards such as ISO/IEC 14443 which has been adopted for biometric passports, credit cards such as American Express's ExpressPay, and many other applications.

Bluetooth 2 will use π/4-DQPSK at its lower rate (2 Mbit/s) and 8-DPSK at its higher rate (3 Mbit/s) when the link between the two devices is sufficiently robust. Bluetooth 1 modulates with Gaussian minimum-shift keying, a binary scheme, so either modulation choice in version 2 will yield a higher data-rate. A similar technology, IEEE 802.15.4 (the wireless standard used by ZigBee) also relies on PSK using two frequency bands: 868–915 MHz with BPSK and at 2.4 GHz with OQPSK.

Both QPSK and 8PSK are widely used in satellite broadcasting. QPSK is still widely used in the streaming of SD satellite channels and some HD channels. High definition programming is delivered almost exclusively in 8PSK due to the higher bitrates of HD video and the high cost of satellite bandwidth. The DVB-S2 standard requires support for both QPSK and 8PSK. The chipsets used in new satellite set top boxes, such as Broadcom's 7000 series support 8PSK and are backward compatible with the older standard.

Historically, voice-band synchronous modems such as the Bell 201, 208, and 209 and the CCITT V.26, V.27, V.29, V.32, and V.34 used PSK.

Binary Phase-shift Keying (BPSK)

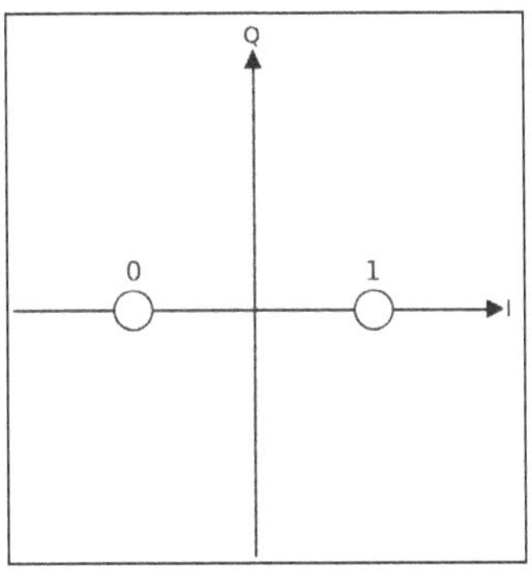

Constellation diagram example for BPSK.

BPSK (also sometimes called PRK, phase reversal keying, or 2PSK) is the simplest form of phase shift keying (PSK). It uses two phases which are separated by 180° and so can also be termed 2-PSK. It does not particularly matter exactly where the constellation points are positioned, and in this figure they are shown on the real axis, at 0° and 180°. Therefore, it handles the highest noise level or distortion before the demodulator reaches an incorrect decision. That makes it the most robust of all the PSKs. It is, however, only able to modulate at 1 bit/symbol and so is unsuitable for high data-rate applications.

In the presence of an arbitrary phase-shift introduced by the communications channel, the demodulator is unable to tell which constellation point is which. As a result, the data is often differentially encoded prior to modulation.

BPSK is functionally equivalent to 2-QAM modulation.

Implementation

The general form for BPSK follows the equation:

$$s_n(t) = \sqrt{\frac{2E_b}{T_b}} \cos(2\pi ft + \pi(1-n)), \quad n = 0,1.$$

This yields two phases, 0 and π. In the specific form, binary data is often conveyed with the following signals:

$$s_0(t) = \sqrt{\frac{2E_b}{T_b}} \cos(2\pi ft + \pi) = -\sqrt{\frac{2E_b}{T_b}} \cos(2\pi ft) \text{ for binary "0"}$$
$$s_1(t) = \sqrt{\frac{2E_b}{T_b}} \cos(2\pi ft) \text{ for binary "1"},$$

where f is the frequency of the base band.

Hence, the signal space can be represented by the single basis function:

$$\phi(t) = \sqrt{\frac{2}{T_b}} \cos(2\pi ft),$$

where 1 is represented by $\sqrt{E_b}\phi(t)$ and 0 is represented by $-\sqrt{E_b}\phi(t)$. This assignment is arbitrary.

This use of this basis function is shown at the end of the next section in a signal timing diagram. The topmost signal is a BPSK-modulated cosine wave that the BPSK modulator would produce. The bit-stream that causes this output is shown above the signal (the other parts of this figure are relevant only to QPSK). After modulation, the base band signal will be moved to the high frequency band by multiplying $\cos(2\pi f_c t)$.

Bit Error Rate

The bit error rate (BER) of BPSK under additive white Gaussian noise (AWGN) can be calculated as:

$$P_b = Q\left(\sqrt{\frac{2E_b}{N_0}}\right) \text{ or } P_e = \frac{1}{2}\operatorname{erfc}\left(\sqrt{\frac{E_b}{N_0}}\right).$$

Quadrature Phase-shift Keying (QPSK)

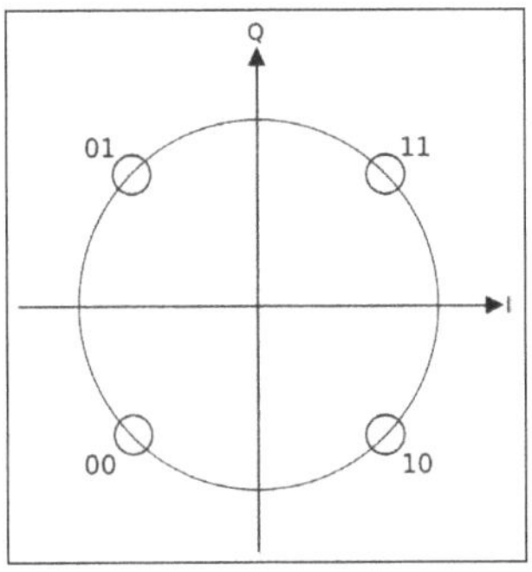

Constellation diagram for QPSK with Gray coding.
Each adjacent symbol only differs by one bit.

Sometimes this is known as quadriphase *PSK*, 4-PSK, or 4-QAM. (Although the root concepts of QPSK and 4-QAM are different, the resulting modulated radio waves are exactly the same). QPSK uses four points on the constellation diagram, equispaced around a circle. With four phases, QPSK can encode two bits per symbol, with Gray coding to minimize the bit error rate (BER) – sometimes misperceived as twice the BER of BPSK.

The mathematical analysis shows that QPSK can be used either to double the data rate compared with a BPSK system while maintaining the same bandwidth of the signal, or to maintain the data-rate *of BPSK* but halving the bandwidth needed. In this latter case, the BER of QPSK is exactly the same as the BER of BPSK – and deciding differently is a common confusion when considering or describing QPSK. The transmitted carrier can undergo numbers of phase changes.

Given that radio communication channels are allocated by agencies such as the Federal Communications Commission giving a prescribed (maximum) bandwidth, the advantage of QPSK over BPSK becomes evident: QPSK transmits twice the data rate in a given bandwidth compared to BPSK - at the same BER. The engineering penalty that is paid is that QPSK transmitters and receivers are more complicated than the ones for BPSK. However, with modern electronics technology, the penalty in cost is very moderate.

As with BPSK, there are phase ambiguity problems at the receiving end, and differentially encoded QPSK is often used in practice.

Implementation

The implementation of QPSK is more general than that of BPSK and also indicates the implementation of higher-order PSK. Writing the symbols in the constellation diagram in terms of the sine and cosine waves used to transmit them:

$$s_n(t) = \sqrt{\frac{2E_s}{T_s}} \cos\left(2\pi f_c t + (2n-1)\frac{\pi}{4}\right), \quad n = 1,2,3,4.$$

This yields the four phases π/4, 3π/4, 5π/4 and 7π/4 as needed.

This results in a two-dimensional signal space with unit basis functions:

$$\phi_1(t) = \sqrt{\frac{2}{T_s}}\cos(2\pi f_c t)$$

$$\phi_2(t) = \sqrt{\frac{2}{T_s}}\sin(2\pi f_c t).$$

The first basis function is used as the in-phase component of the signal and the second as the quadrature component of the signal.

Hence, the signal constellation consists of the signal-space 4 points:

$$\left(\pm\sqrt{\frac{E_s}{2}}, \pm\sqrt{\frac{E_s}{2}}\right).$$

The factors of 1/2 indicate that the total power is split equally between the two carriers.

Comparing these basis functions with that for BPSK shows clearly how QPSK can be viewed as two independent BPSK signals. Note that the signal-space points for BPSK do not need to split the symbol (bit) energy over the two carriers in the scheme shown in the BPSK constellation diagram.

QPSK systems can be implemented in a number of ways. An illustration of the major components of the transmitter and receiver structure are shown below.

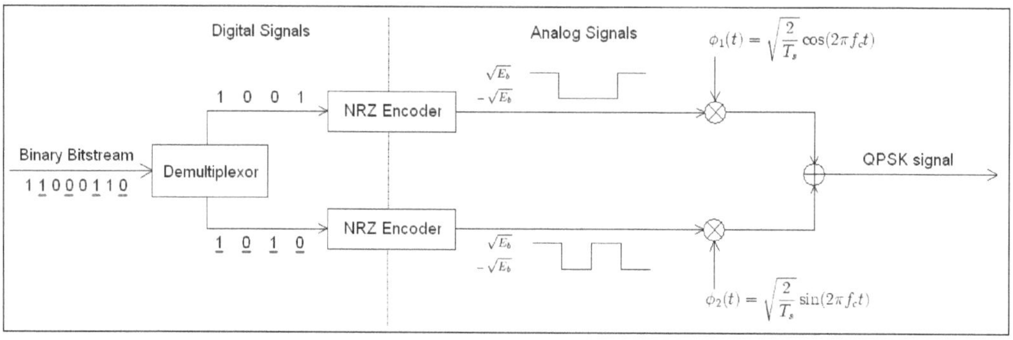

Conceptual transmitter structure for QPSK. The binary data stream is split into the in-phase and quadrature-phase components. These are then separately modulated onto two orthogonal basis functions. In this implementation, two sinusoids are used. Afterwards, the two signals are superimposed, and the resulting signal is the QPSK signal. Note the use of polar non-return-to-zero encoding. These encoders can be placed before for binary data source, but have been placed after to illustrate the conceptual difference between digital and analog signals involved with digital modulation.

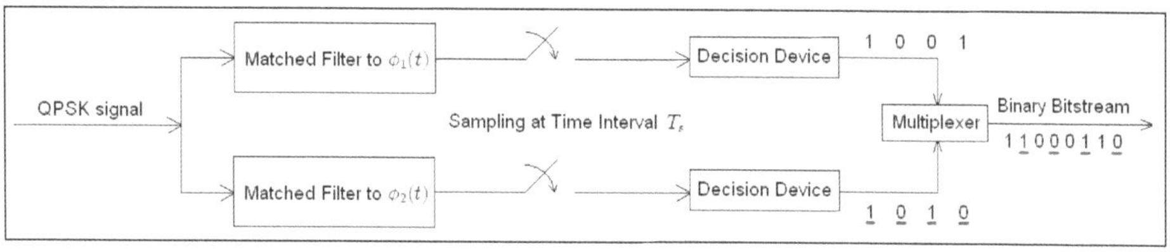

Receiver structure for QPSK. The matched filters can be replaced with correlators. Each detection device uses a reference threshold value to determine whether a 1 or 0 is detected.

Probability of Error

Although QPSK can be viewed as a quaternary modulation, it is easier to see it as two independently modulated quadrature carriers. With this interpretation, the even (or odd) bits are used to modulate the in-phase component of the carrier, while the odd (or even) bits are used to modulate the quadrature-phase component of the carrier. BPSK is used on both carriers and they can be independently demodulated.

As a result, the probability of bit-error for QPSK is the same as for BPSK:

$$P_b = Q\left(\sqrt{\frac{2E_b}{N_0}}\right)$$

However, in order to achieve the same bit-error probability as BPSK, QPSK uses twice the power (since two bits are transmitted simultaneously).

The symbol error rate is given by:

$$\begin{aligned} P_s &= 1-(1-P_b)^2 \\ &= 2Q\left(\sqrt{\frac{E_s}{N_0}}\right) - \left[Q\left(\sqrt{\frac{E_s}{N_0}}\right)\right]^2. \end{aligned}$$

If the signal-to-noise ratio is high (as is necessary for practical QPSK systems) the probability of symbol error may be approximated:

$$P_s \approx 2Q\left(\sqrt{\frac{E_s}{N_0}}\right) = \operatorname{erfc}\left(\sqrt{\frac{E_s}{2N_0}}\right) = \operatorname{erfc}\left(\sqrt{\frac{E_b}{N_0}}\right).$$

The modulated signal is shown below for a short segment of a random binary data-stream. The two carrier waves are a cosine wave and a sine wave, as indicated by the signal-space analysis above. Here, the odd-numbered bits have been assigned to the in-phase component and the even-numbered bits to the quadrature component (taking the first bit as number 1). The total signal – the sum of the two components – is shown at the bottom. Jumps in phase can be seen as the PSK changes the phase on each component at the start of each bit-period. The topmost waveform alone matches the description given for BPSK above.

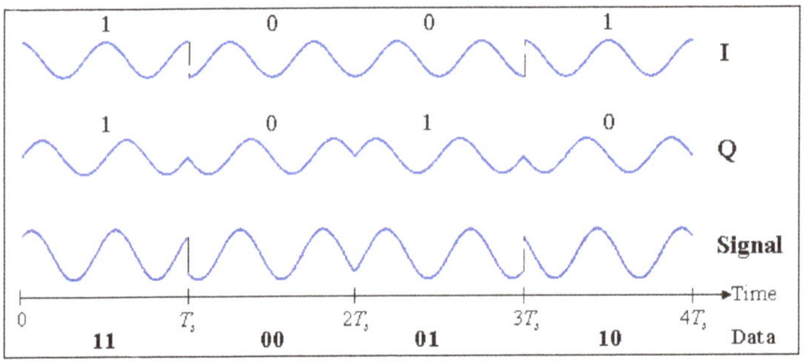

Timing diagram for QPSK. The binary data stream is shown beneath the time axis. The two signal components with their bit assignments are shown at the top, and the total combined signal at the bottom. Note the abrupt changes in phase at some of the bit-period boundaries.

The binary data that is conveyed by this waveform is: 11000110.

- The odd bits, highlighted here, contribute to the in-phase component: 11000110.
- The even bits, highlighted here, contribute to the quadrature-phase component: 11000110.

Variants

Offset QPSK (OQPSK)

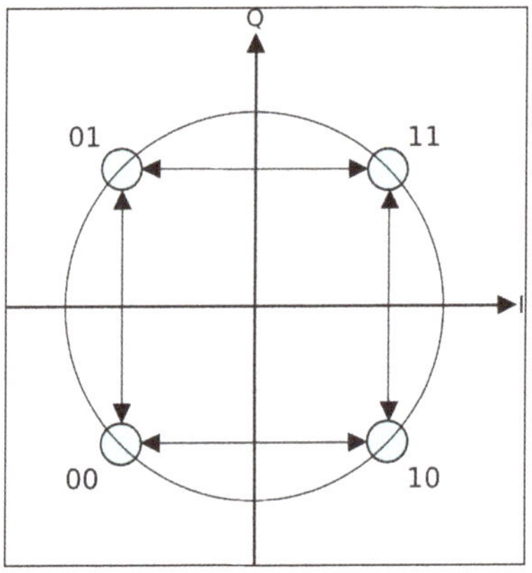

Signal doesn't pass through the origin, because only one bit of the symbol is changed at a time.

Offset quadrature phase-shift keying (OQPSK) is a variant of phase-shift keying modulation using four different values of the phase to transmit. It is sometimes called staggered quadrature phase-shift keying (SQPSK).

Taking four values of the phase (two bits) at a time to construct a QPSK symbol can allow the phase of the signal to jump by as much as 180° at a time. When the signal is low-pass filtered (as is typical in a transmitter), these phase-shifts result in large amplitude fluctuations, an undesirable quality in communication systems. By offsetting the timing of the odd and even bits by one bit-period, or half a symbol-period, the in-phase and quadrature components will never change at the same time. In the constellation diagram shown on the right, it can be seen that this will limit the phase-shift to no more than 90° at a time. This yields much lower amplitude fluctuations than non-offset QPSK and is sometimes preferred in practice.

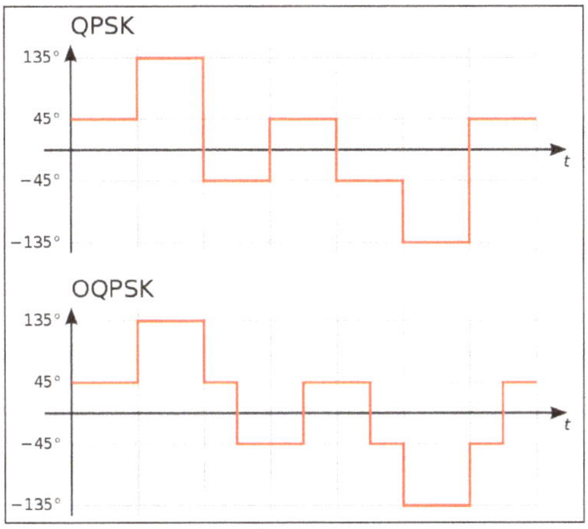

Difference of the phase between QPSK and OQPSK.

The picture shows the difference in the behavior of the phase between ordinary QPSK and OQPSK. It can be seen that in the first plot the phase can change by 180° at once, while in OQPSK the changes are never greater than 90°.

The modulated signal is shown below for a short segment of a random binary data-stream. Note the half symbol-period offset between the two component waves. The sudden phase-shifts occur about twice as often as for QPSK (since the signals no longer change together), but they are less severe. In other words, the magnitude of jumps is smaller in OQPSK when compared to QPSK.

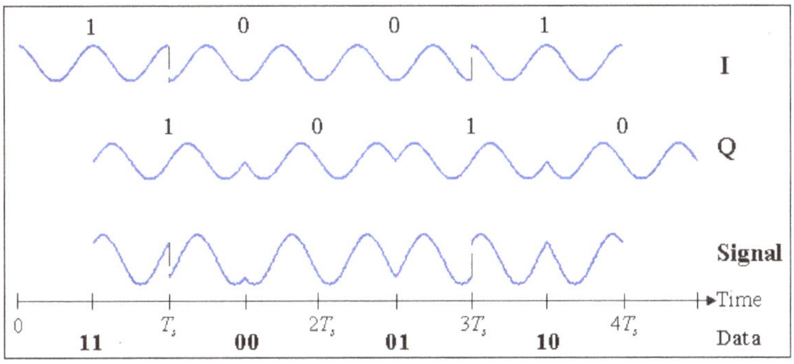

Timing diagram for offset-QPSK. The binary data stream is shown beneath the time axis. The two signal components with their bit assignments are shown the top and the total, combined signal at the bottom. Note the half-period offset between the two signal components.

π/4-QPSK

This variant of QPSK uses two identical constellations which are rotated by 45° ($\pi/4$ radians, hence the name) with respect to one another. Usually, either the even or odd symbols are used to select points from one of the constellations and the other symbols select points from the other constellation. This also reduces the phase-shifts from a maximum of 180°, but only to a maximum of 135° and so the amplitude fluctuations of $\pi/4$ -QPSK are between OQPSK and non-offset QPSK.

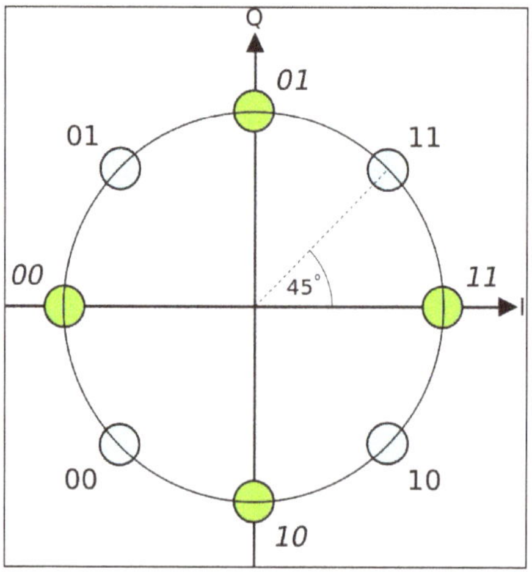

Dual constellation diagram for π/4-QPSK. This shows the two separate constellations with identical Gray coding but rotated by 45° with respect to each other.

One property this modulation scheme possesses is that if the modulated signal is represented in the complex domain, transitions between symbols never pass through 0. In other words, the signal does not pass through the origin. This lowers the dynamical range of fluctuations in the signal which is desirable when engineering communications signals.

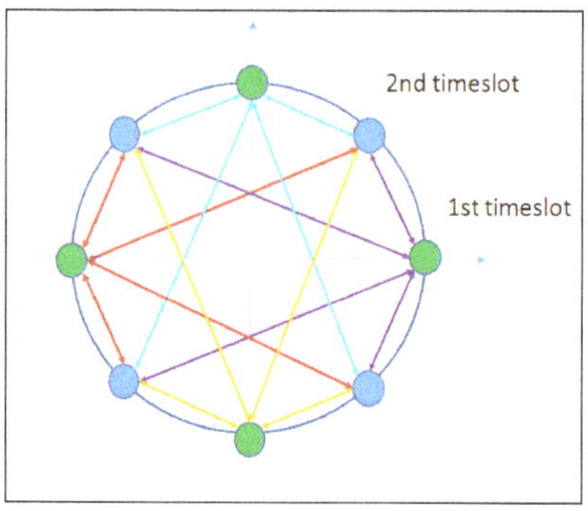

Transition scheme of the modulation symbols of the π/4-QPSK (signal constellation). No zero crossings.

On the other hand, $\pi/4$ -QPSK lends itself to easy demodulation and has been adopted for use in, for example, TDMA cellular telephone systems.

The modulated signal is shown below for a short segment of a random binary data-stream. The construction is the same as above for ordinary QPSK. Successive symbols are taken from the two constellations shown in the diagram. Thus, the first symbol (1 1) is taken from the "blue" constellation and the second symbol (0 0) is taken from the "green" constellation. Note that magnitudes of the two component waves change as they switch between constellations, but the total signal's magnitude remains constant (constant envelope). The phase-shifts are between those of the two previous timing-diagrams.

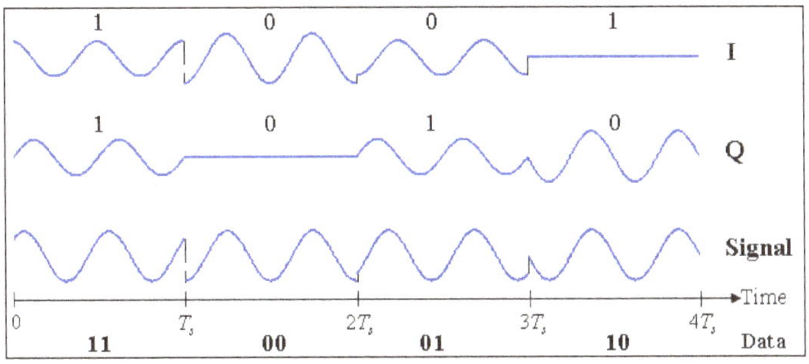

Timing diagram for π/4-QPSK. The binary data stream is shown beneath the time axis. The two signal components with their bit assignments are shown the top and the total, combined signal at the bottom. Note that successive symbols are taken alternately from the two constellations, starting with the "blue" one.

SOQPSK

The license-free shaped-offset QPSK (SOQPSK) is interoperable with Feher-patented QPSK (FQPSK), in the sense that an integrate-and-dump offset QPSK detector produces the same output no matter which kind of transmitter is used.

These modulations carefully shape the I and Q waveforms such that they change very smoothly, and the signal stays constant-amplitude even during signal transitions. (Rather than traveling instantly from one symbol to another, or even linearly, it travels smoothly around the constant-amplitude circle from one symbol to the next). The standard description of SOQPSK-TG involves ternary symbols.

DPQPSK

Dual-polarization quadrature phase shift keying (DPQPSK) or dual-polarization QPSK - involves the polarization multiplexing of two different QPSK signals, thus improving the spectral efficiency by a factor of 2. This is a cost-effective alternative to utilizing 16-PSK, instead of QPSK to double the spectral efficiency.

Higher-order PSK

Any number of phases may be used to construct a PSK constellation but 8-PSK is usually the highest order PSK constellation deployed. With more than 8 phases, the error-rate becomes too high and there are better, though more complex, modulations available such as quadrature amplitude modulation (QAM). Although any number of phases may be used, the fact that the constellation must usually deal with binary data means that the number of symbols is usually a power of 2 to allow an integer number of bits per symbol.

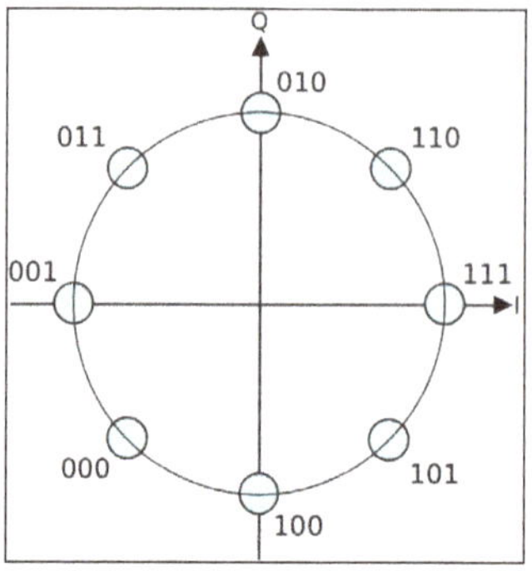

Constellation diagram for 8-PSK with Gray coding.

Bit Error Rate

For the general M-PSK there is no simple expression for the symbol-error probability if $M > 4$. Unfortunately, it can only be obtained from:

$$P_s = 1 - \int_{-\pi/M}^{\pi/M} p_{\theta_r}(\theta_r)\, d\theta_r,$$

Where,

$$p_{\theta_r}(\theta_r) = \frac{1}{2\pi} e^{-2\gamma_s \sin^2 \theta_r} \int_0^{\infty} V e^{-\frac{1}{2}\left(V - 2\sqrt{\gamma_s}\cos\theta_r\right)^2} dV,$$

$$V = \sqrt{r_1^2 + r_2^2},$$

$$\theta_r = \tan^{-1}\left(\frac{r_2}{r_1}\right),$$

$$\gamma_s = \frac{E_s}{N_0}$$

and $r_1 \sim N\left(\sqrt{E_s}, \frac{1}{2}N_0\right)$ and $r_2 \sim N\left(0, \frac{1}{2}N_0\right)$ are each Gaussian random variables.

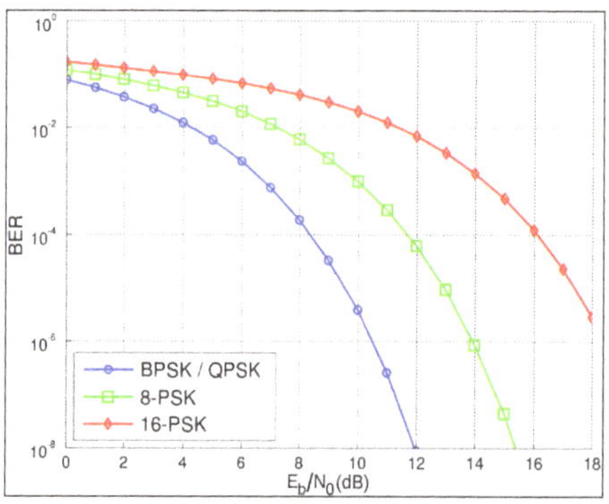

Bit-error rate curves for BPSK, QPSK, 8-PSK and 16-PSK, additive white Gaussian noise channel.

This may be approximated for high M and high E_b / N_0 by:

$$P_s \approx 2Q\left(\sqrt{2\gamma_s}\sin\frac{\pi}{M}\right).$$

The bit-error probability for M-PSK can only be determined exactly once the bit-mapping is known. However, when Gray coding is used, the most probable error from one symbol to the next produces only a single bit-error and,

$$P_b \approx \frac{1}{k}P_s$$

(Using Gray coding allows us to approximate the Lee distance of the errors as the Hamming distance of the errors in the decoded bitstream, which is easier to implement in hardware).

The graph on the left compares the bit-error rates of BPSK, QPSK (which are the same, 8-PSK and 16-PSK. It is seen that higher-order modulations exhibit higher error-rates; in exchange however they deliver a higher raw data-rate.

Bounds on the error rates of various digital modulation schemes can be computed with application of the union bound to the signal constellation.

Differential Phase-shift Keying (DPSK)

Differential Encoding

Differential phase shift keying (DPSK) is a common form of phase modulation that conveys data by changing the phase of the carrier wave. As mentioned for BPSK and QPSK there is an ambiguity of phase if the constellation is rotated by some effect in the communications channel through which the signal passes. This problem can be overcome by using the data to change rather than set the phase.

For example, in differentially encoded BPSK a binary "1" may be transmitted by adding 180° to the current phase and a binary "0" by adding 0° to the current phase. Another variant of DPSK is Symmetric Differential Phase Shift keying, SDPSK, where encoding would be +90° for a "1" and −90° for a "0".

In differentially encoded QPSK (DQPSK), the phase-shifts are 0°, 90°, 180°, −90° corresponding to data "00", "01", "11", "10". This kind of encoding may be demodulated in the same way as for non-differential PSK but the phase ambiguities can be ignored. Thus, each received symbol is demodulated to one of the M points in the constellation and a comparator then computes the difference in phase between this received signal and the preceding one. The difference encodes the data as described above. Symmetric differential quadrature phase shift keying (SDQPSK) is like DQPSK, but encoding is symmetric, using phase shift values of −135°, −45°, +45° and +135°.

The modulated signal is shown below for both DBPSK and DQPSK as described above. In the figure, it is assumed that the signal starts with zero phase, and so there is a phase shift in both signals at $t = 0$.

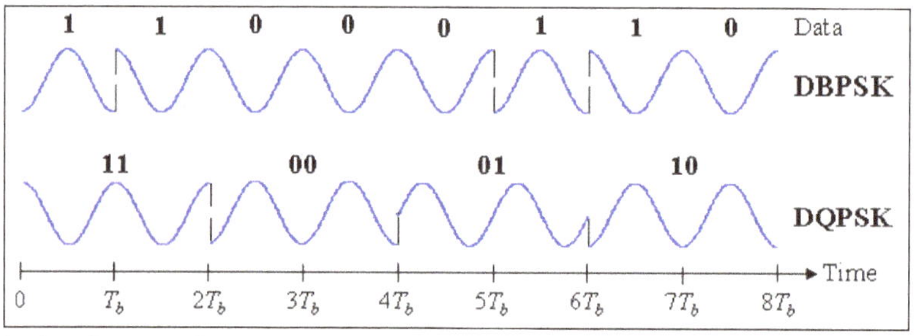

Timing diagram for DBPSK and DQPSK. The binary data stream is above the DBPSK signal. The individual bits of the DBPSK signal are grouped into pairs for the DQPSK signal, which only changes every $T_s = 2T_b$.

Analysis shows that differential encoding approximately doubles the error rate compared to ordinary *M*-PSK but this may be overcome by only a small increase in E_b / N_0. Furthermore, this analysis (and the graphical results below) are based on a system in which the only corruption is additive white Gaussian noise (AWGN). However, there will also be a physical channel between the transmitter and receiver in the communication system. This channel will, in general, introduce an unknown phase-shift to the PSK signal; in these cases the differential schemes can yield a *better* error-rate than the ordinary schemes which rely on precise phase information.

One of the most popular applications of DPSK is the Bluetooth standard where $\pi / 4$ -DQPSK and 8-DPSK were implemented.

Demodulation

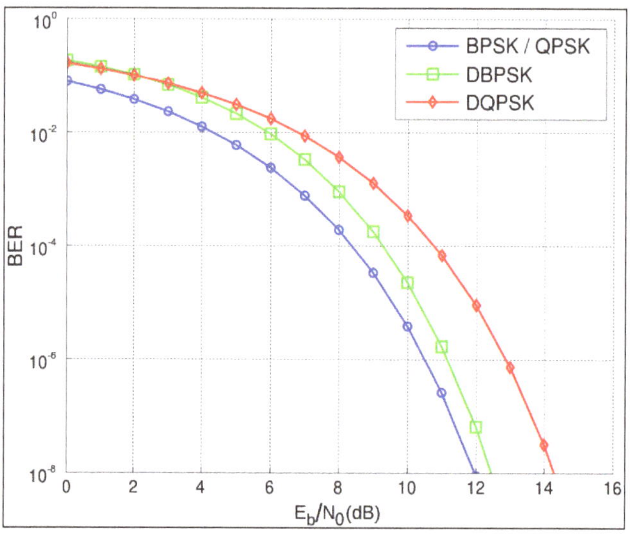

BER comparison between DBPSK, DQPSK and their non-differential forms using Gray coding and operating in white noise.

For a signal that has been differentially encoded, there is an obvious alternative method of demodulation. Instead of demodulating as usual and ignoring carrier-phase ambiguity, the phase between two successive received symbols is compared and used to determine what the data must have been. When differential encoding is used in this manner, the scheme is known as differential phase-shift keying (DPSK). Note that this is subtly different from just differentially encoded PSK since, upon reception, the received symbols are *not* decoded one-by-one to constellation points but are instead compared directly to one another.

Call the received symbol in the kth timeslot r_k and let it have phase ϕ_k. Assume without loss of generality that the phase of the carrier wave is zero. Denote the additive white Gaussian noise (AWGN) term as n_k. Then:

$$r_k = \sqrt{E_s} e^{j\phi_k} + n_k.$$

The decision variable for the $k-1^{\text{th}}$ symbol and the kth symbol is the phase difference between r_k and r_{k-1}. That is, if r_k is projected onto r_{k-1}, the decision is taken on the phase of the resultant complex number:

$$r_k r_{k-1}^* = E_s e^{j(\varphi_k - \varphi_{k-1})} + \sqrt{E_s} e^{j\varphi_k} n_{k-1}^* + \sqrt{E_s} e^{-j\varphi_{k-1}} n_k + n_k n_{k-1}^*$$

where superscript * denotes complex conjugation. In the absence of noise, the phase of this is $\phi_k - \phi_{k-1}$, the phase-shift between the two received signals which can be used to determine the data transmitted.

The probability of error for DPSK is difficult to calculate in general, but, in the case of DBPSK it is:

$$P_b = \frac{1}{2} e^{-\frac{E_b}{N_0}},$$

which, when numerically evaluated, is only slightly worse than ordinary BPSK, particularly at higher E_b / N_0 values.

Using DPSK avoids the need for possibly complex carrier-recovery schemes to provide an accurate phase estimate and can be an attractive alternative to ordinary PSK.

In optical communications, the data can be modulated onto the phase of a laser in a differential way. The modulation is a laser which emits a continuous wave, and a Mach–Zehnder modulator which receives electrical binary data. For the case of BPSK, the laser transmits the field unchanged for binary '1', and with reverse polarity for '0'. The demodulator consists of a delay line interferometer which delays one bit, so two bits can be compared at one time. In further processing, a photodiode is used to transform the optical field into an electric current, so the information is changed back into its original state.

The bit-error rates of DBPSK and DQPSK are compared to their non-differential counterparts in the graph to the right. The loss for using DBPSK is small enough compared to the complexity reduction that it is often used in communications systems that would otherwise use BPSK. For DQPSK though, the loss in performance compared to ordinary QPSK is larger and the system designer must balance this against the reduction in complexity.

Example: Differentially encoded BPSK.

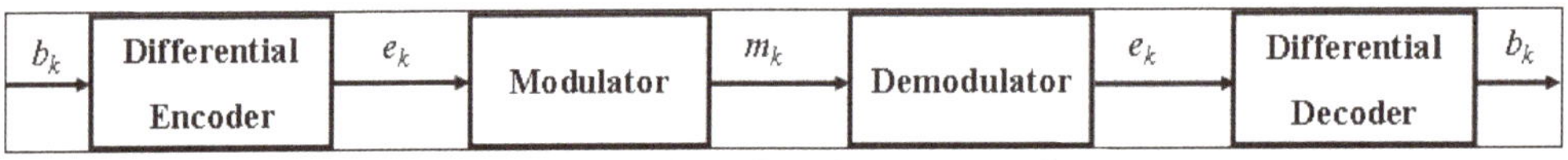

Differential encoding/decoding system diagram.

At the kth time-slot call the bit to be modulated b_k, the differentially encoded bit e_k and the resulting modulated signal $m_k(t)$. Assume that the constellation diagram positions the symbols at ±1 (which is BPSK). The differential encoder produces:

$$e_k = e_{k-1} \oplus b_k$$

where $\oplus$ indicates binary or modulo-2 addition.

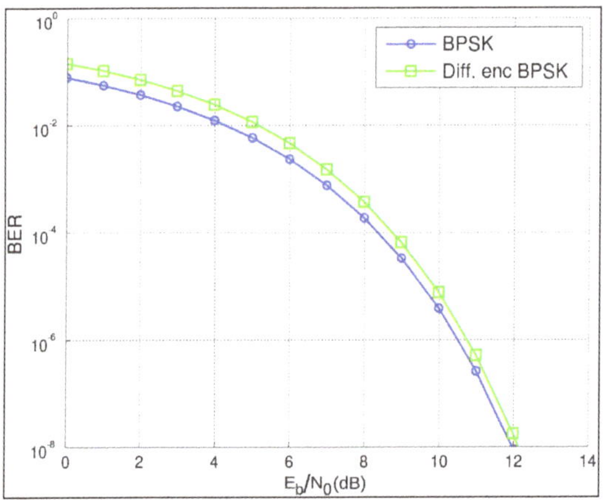

BER comparison between BPSK and differentially encoded BPSK with Gray coding operating in white noise.

So e_k only changes state (from binary "0" to binary "1" or from binary "1" to binary "0") if is a binary "1". Otherwise it remains in its previous state. This is the description of differentially encoded BPSK given above.

The received signal is demodulated to yield $e_k = \pm 1$ and then the differential decoder reverses the encoding procedure and produces:

$$b_k = e_k \oplus e_{k-1},$$

since binary subtraction is the same as binary addition.

Therefore, $b_k = 1$ if e_k and e_{k-1} differ and $b_k = 0$ if they are the same. Hence, if both e_k

and e_{k-1} are *inverted,* b_k will still be decoded correctly. Thus, the 180° phase ambiguity does not matter.

Differential schemes for other PSK modulations may be devised along similar lines. The waveforms for DPSK are the same as for differentially encoded PSK given above since the only change between the two schemes is at the receiver.

The BER curve for this example is compared to ordinary BPSK on the right. As mentioned above, whilst the error rate is approximately doubled, the increase needed in E_b / N_0 to overcome this is small. The increase in E_b / N_0 required to overcome differential modulation in coded systems, however, is larger – typically about 3 dB. The performance degradation is a result of noncoherent transmission – in this case it refers to the fact that tracking of the phase is completely ignored.

Mutual Information with Additive White Gaussian Noise

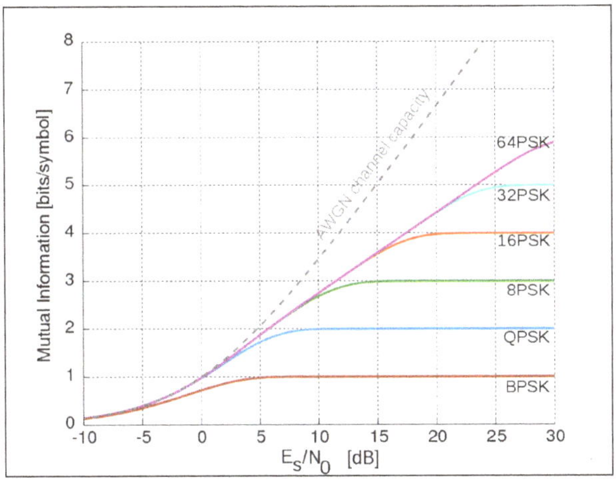

Mutual information of PSK over the AWGN channel.

The mutual information of PSK can be evaluated in additive Gaussian noise by numerical integration of its definition. The curves of mutual information saturate to the number of bits carried by each symbol in the limit of infinite signal to noise ratio E_b / N_0. On the contrary, in the limit of small signal to noise ratios the mutual information approaches the AWGN channel capacity, which is the supremum among all possible choices of symbol statistical distributions.

At intermediate values of signal to noise ratios the mutual information (MI) is well approximated by:

$$\mathrm{MI} \simeq \log_2\left(\sqrt{\frac{4\pi}{e}\frac{E_s}{N_0}}\right).$$

The mutual information of PSK over the AWGN channel is generally farther to the AWGN channel capacity than QAM modulation formats.

References

- Digital-communication-introduction-basic-components-how-signal-process-works-and-advan-tages: electricalfundablog.com, Retrieved 25 August, 2020
- Digital-communication-quick-guide, digital-communication: tutorialspoint.com, Retrieved 04 March, 2020
- David R. Smith, "Digital Transmission Systems", Kluwer International Publishers, 2003, ISBN 1-4020-7587-1
- Nelson, T.; Perrins, E.; Rice, M. (2005). "Common detectors for shaped offset QPSK (SOQPSK) and Feher-patented QPSK (FQPSK)". GLOBECOM '05. IEEE Global Telecommunications Conference, 2005. pp. 5 pp. doi:10.1109/GLOCOM.2005.1578470. ISBN 0-7803-9414-3. ISBN 0-7803-9414-3
- Pulse-code-modulation-and-demodulation: elprocus.com, Retrieved 14 July, 2020

Permissions

All chapters in this book are published with permission under the Creative Commons Attribution Share Alike License or equivalent. Every chapter published in this book has been scrutinized by our experts. Their significance has been extensively debated. The topics covered herein carry significant information for a comprehensive understanding. They may even be implemented as practical applications or may be referred to as a beginning point for further studies.

We would like to thank the editorial team for lending their expertise to make the book truly unique. They have played a crucial role in the development of this book. Without their invaluable contributions this book wouldn't have been possible. They have made vital efforts to compile up to date information on the varied aspects of this subject to make this book a valuable addition to the collection of many professionals and students.

This book was conceptualized with the vision of imparting up-to-date and integrated information in this field. To ensure the same, a matchless editorial board was set up. Every individual on the board went through rigorous rounds of assessment to prove their worth. After which they invested a large part of their time researching and compiling the most relevant data for our readers.

The editorial board has been involved in producing this book since its inception. They have spent rigorous hours researching and exploring the diverse topics which have resulted in the successful publishing of this book. They have passed on their knowledge of decades through this book. To expedite this challenging task, the publisher supported the team at every step. A small team of assistant editors was also appointed to further simplify the editing procedure and attain best results for the readers.

Apart from the editorial board, the designing team has also invested a significant amount of their time in understanding the subject and creating the most relevant covers. They scrutinized every image to scout for the most suitable representation of the subject and create an appropriate cover for the book.

The publishing team has been an ardent support to the editorial, designing and production team. Their endless efforts to recruit the best for this project, has resulted in the accomplishment of this book. They are a veteran in the field of academics and their pool of knowledge is as vast as their experience in printing. Their expertise and guidance has proved useful at every step. Their uncompromising quality standards have made this book an exceptional effort. Their encouragement from time to time has been an inspiration for everyone.

The publisher and the editorial board hope that this book will prove to be a valuable piece of knowledge for students, practitioners and scholars across the globe.

Index

www.ingramcontent.com/pod-product-compliance
Lightning Source LLC
LaVergne TN
LVHW080630210926
848485LV00003B/237